Electric Revolution: Shifting Gears Towards a Sustainable Future

Ravi

Table of Contents

**Chapter 1: Introduction to PEV Adoption in the US and Prior Research:
What Are PEVs and Where Do We Stand Today?**

The automotive industry is on the precipice of a monumental paradigm shift with

the development of promising vehicle technologies aimed at reducing the overall

environmental impact of vehicular transportation. New vehicle technologies involving

plug-in electric vehicles (PEVs) offer a promising pathway to rapid decarbonization of

the transportation industry, the largest contributor to anthropogenic greenhouse gas

emissions in the United States (US) [1], provided they are charged on low-carbon energy

sources. Despite the environmental benefits, innovative features, and potential wide range

of government incentives to increase PEV adoption, sales are still low compared to sales

of conventional Internal Combustion Engine (ICE) vehicles [2]–[4]. While the market

share of PEVs has continued to increase in the US, they remain a relatively niche offer; in

2023 US PEV sales comprised just over 9% of total new vehicle sales [5]. The

transportation industry has also become the largest emitter of GHG emissions in the US

in recent years, with a little over 1.8B tons of CO_2 equivalent emitted in 2021 [6].

Addressing emissions within the transportation sector is critical, and gaining further

consumer adoption of PEVs is the path forward towards this future.

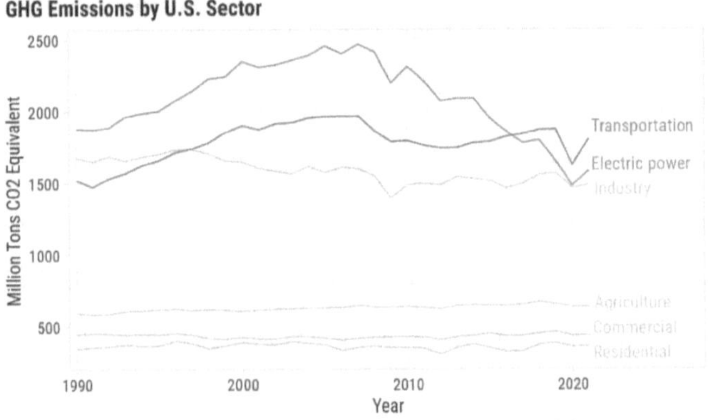

Figure 1-1: US GHG Emissions by Sector.
The transportation sector is the largest emitter of GHG emissions and has continued to increase in 2021 despite a brief decrease during COVID [6], [7].

As frequently observed with the introduction of innovative technologies, consumers have many questions and uncertainties surrounding PEVs. Personal vehicles are a critical component to overall mobility and daily life in the US, and in 2022 was the 2nd largest consumer expenditure for US households behind housing [8]. Consumer reservations towards PEVs span many different technical and socio-economic aspects of the technology, including utilization, longevity, and accessibility and may reflect incorrect or outdated perceptions. This significance combined with lack of knowledge and potential misperceptions about PEVs creates an ambiguous future for consumers around adoption of this new innovation.

In order to deliver the environmental benefits and other gains from this new vehicle technology, research is essential to understand how consumers assess and accept PEVs. If PEVs are going to make the transition to a mass-market vehicle, then further work is needed to understand how to propel forward consumer acceptance of these vehicles. This book focuses on opportunities to gain further consumer acceptance of PEVs studying specific mechanisms available to the industry, policymakers and stakeholders. These studies focus on alternative powertrains, including hybrid (HEV), plug-in hybrid (PHEV), battery electric (BEV), and fuel cell vehicles (FCEV) with an emphasis on PHEVs and BEVs (PEVs). All studies have a consumer element with different inputs that assist in consideration of PEVs.

We achieved the stated objectives through the execution of three research studies:

1. The impact of short exposure PEV experience on consumer acceptance
2. PEV financial incentives based on consumer preferences
3. Investigation of resale value of PEVs vs. ICEs

These studies together connect back to PEV adoption by contributing insight into the various phases along the innovation diffusion curve. Utilizing Rogers' Diffusion of Innovation curve built from the Bass Model, there are many phases in consumer adoption of an innovative technology [9]. These three studies aim to bring consumer insight along different stages of PEV adoption in order to create a holistic book on the progression of PEVs in the US and how consumers adopt new technologies. In addition, all three studies aim to highlight ways to prioritize equity measures in this innovation diffusion, as diffusion of innovations typically broaden the gap between lower- and upper-income groups and moving this technology forward will require access across all parts of the US population.

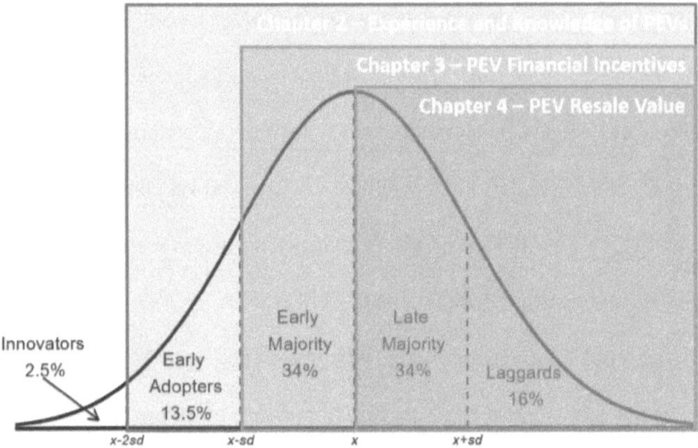

Figure 1-2: Innovation Adoption Curve Overlaid with book Research Chapters and Consumer Impact.
Figure is based on Rogers adoption curve and classifications [9] and recreated by author.

These chapters deliver wider contributions to the research community. First, the results expand on prior works by investigating novel, more time-effective methods by which to increase PEV acceptance. Second, results come from unique data sources that provide consumer-specific data supplied to various modeling techniques. Finally, these studies provide a comprehensive review of important consumer relevant factors meant to bridge the gap between current innovations and consumer approval while considering equity of innovation diffusion as well.

1.1 Background

The electrified vehicle market has emerged as a focal development effort for the future of transportation. Electrified vehicle technologies take various forms in today's automotive market:

Table 1-1: Electrified Vehicle Technologies and Their Attributes.

Electrified Vehicle Technology	Description	Electric Range	Environmental Benefits
Mild Hybrid	Small electric motor (48V) system attached to ICE engine	None	Limited
Hybrid	Larger electric motor attached to ICE engine which can power the vehicle for short distances and at low speeds	<1 mi	🍃
Plug-In Hybrid	Large electric motor attached to ICE engine with large battery pack for electric-only driving; can be plugged in	20-40 mi	🍃 🍃
Battery Electric	Large electric motor and battery pack for electric-only driving; can be plugged in	150-400 mi	🍃 🍃 🍃
Fuel Cell Electric	Large electric motor and battery pack for electric-only driving; electricity generated from a fuel cell powered by hydrogen	300+ mi	🍃 🍃 🍃

Source: DOE Alternative Fuels Data Center [10], [11].

Research has created a strong foundation for diffusion models of technological innovations. The "Bass Model" is the most well-known innovation adoption model and is used across many different fields of study to represent the diffusion of technology across a population, as shown in Figure 1-3 [12]. Extensions and variations of the Bass Model are frequent, as with Rogers' version overlaying information on different groupings of consumers in different phases of the Bass Model [9]. These diffusion models can also integrate stated preferences and reveal characteristics of groupings of adopters, like those known as "early/late majority adopters" in the Bass model, who are much more risk averse and less willing to purchase an innovation so different from the dominant design [13], [14]. Rogers' theory also includes equity issues in diffusion of innovation research, both the lack of research addressing and the hypothesis that diffusion of innovations creates a wider gap between socioeconomic statuses. The various studies in this book extend this foundational theory by offering observations of consumers in different phases of the innovation diffusion curve, as well as analyzing equity strategies.

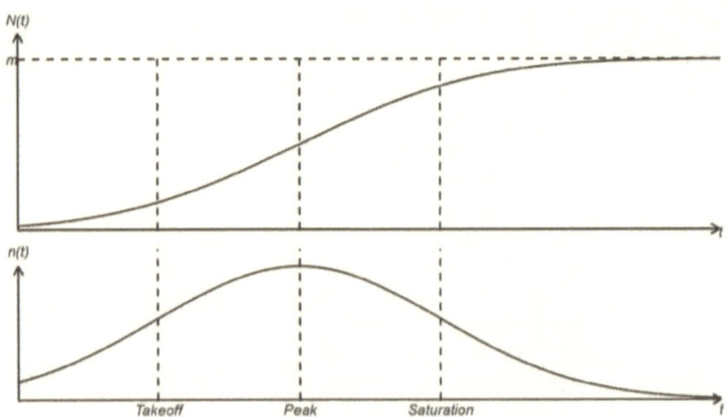

Figure 1-3: Bass Model Cumulative and Noncumulative Adopter Curve.
Figure recreated by author based on Norton and Rogers [12].

Some research has also further investigated the psychological constructs by which

individuals assess and accept new technologies. The theory of planned behavior (TPB) by

Ajzen explains the factors present that influence consumer reception and adoption of new

technologies. According to TPB, the main determining factors of behavioral intention are

attitudes, which are influenced by knowledge and experience, subjective norms that the

consumer believes is acceptable by society, and the perceived control of the behavior or

perceived ability to execute a behavior. An illustration of TPB is found in Figure 1-4. For

the purposes of diffusion of technology research like PEV adoption, consumer acceptance

of technology is considered an intention to adopt, use, or support its development [15].

PEV research has provided evidence that shows the relevance of one or all of the TPB

factors in decision making and adoption of EVs [16]–[21]. Attitudes and perceptions of

technology have been shown to impact consumer behavior, and understanding of a

technology through knowledge and experience acquisition leads to less perceived risk

and more favorable intentions. Higher comprehension of the technology also improves perceptions of behavior impact when it comes to new technology [21]. TPB is an important theoretical construct of the behavioral intentions of individuals when faced with consideration and adoption of a new technology such as PEVs.

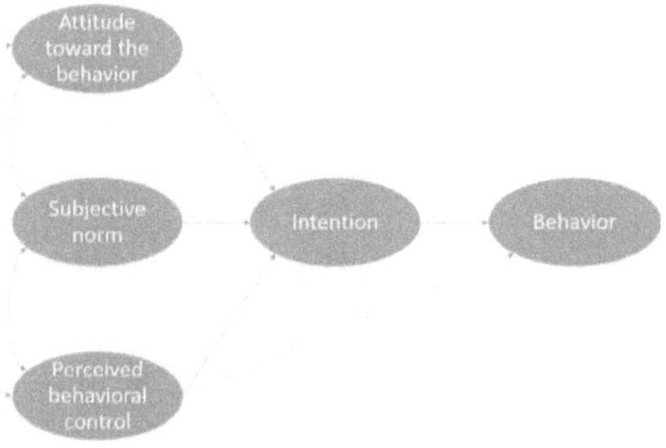

Figure 1-4: Theory of Planned Behavior Sequencing.
Figure recreated by author from TPB illustration Ajzen [15].

As a further extension of TPB, frameworks for the diffusion of innovation have been developed and explored in order to theoretically define the process and contributors to adoption. In a well-known framework developed by Rogers, there are five sequential stages in innovation adoption: (1) gaining knowledge of an innovation, (2) forming an attitude, (3) deciding to adopt or reject it, (4) executing the decision, and (5) confirming the decision [9], [22], [23]. Previous experiences, existing needs/problems, innovativeness and social norms influence consumers to begin this adoption process. Figure 1-5 shows the progression of this innovation diffusion framework and the importance of the individual and their perceptions of characteristics of the innovation as critical inputs into the Knowledge and Persuasion (attitude) steps. Knowledge and experience emerge as important factors before and during Rogers' innovation adoption framework. Rogers also connects his framework to the Bass Model, supposing that 10–

25% of the population rapidly move to adoption of an innovation followed by the

remainder of the population, forming the S-shaped curve found in the Bass Model [9].

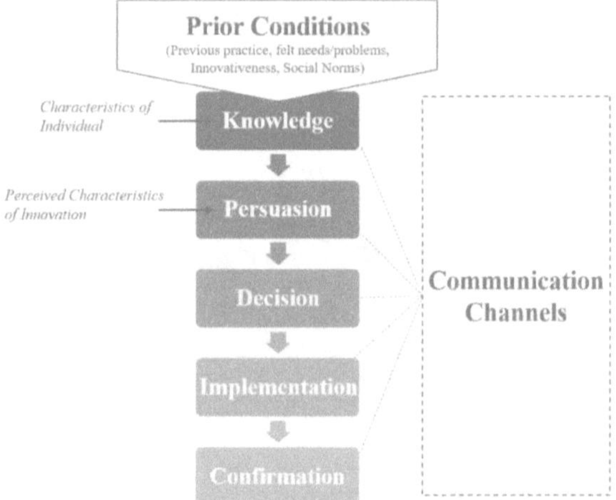

Figure 1-5: Model of Five Stages in the Innovation-Decision Process.
Figure created by author based on Rogers [9].

While considering the rate of adoption, it is also important to highlight the

perceived attributes of innovations and their impact of rate of acceptance of innovations.

Rogers estimates that between 49%-87% of variance in rate of adoption of innovations is

explained by these attributes [9]. Diffusion of innovations is an uncertainty reduction

process and, as such, electric vehicles have some advantages and risks when analyzing

through this model. While there are many similarities to the established technology on a

fundamental level, perceived differences in basic operational, technical and

socioeconomic levels present challenges when looking to accelerate adoption.

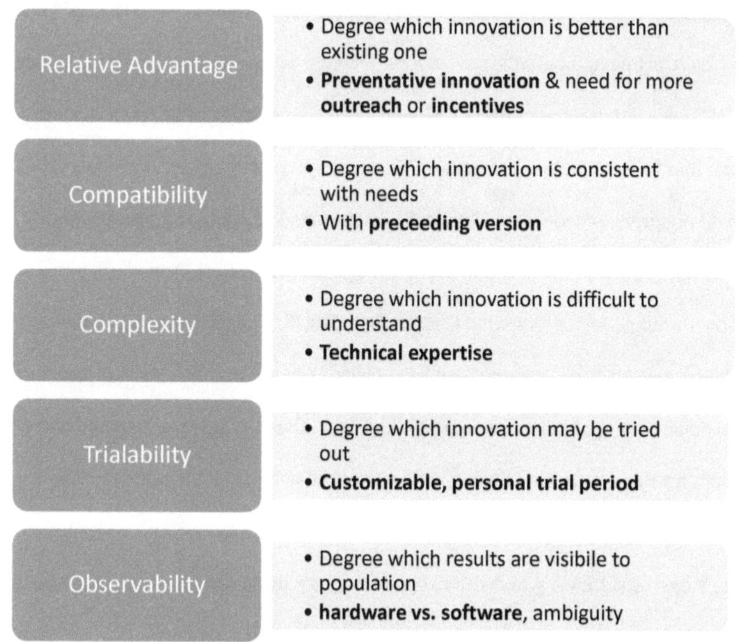

Figure 1-6: Perceived Attributes of Adoption with Relevant Subtopics for PEV Adoption. Figure created by author based on Rogers [9].

This book translates these theories into empirical, applied studies by which to draw further consumer insights and progress diffusion of this innovation. Utilizing this theoretical basis is appropriate for diffusion of electric vehicles and the various consumer dynamics studied for many reasons. First, electric vehicles serve as an example of an innovation as human behavior and theory defines states an idea, practice or object does not need to be "objectively" new or not. Rogers provides a succinct explanation for the definition of an innovation: "If an idea seems new for the individual, it is an innovation" [9]. Given the newness and distinction between electric and conventional vehicles, as well as the new knowledge and uncertainty associated with this new technology, electric vehicles fit into innovation diffusion theory as posited by Rogers. Next, Rogers's theory provides structure and definition to various adoption categories and their traits. Chapter 2 of this book examines the impact of experience and knowledge of PEVs, providing insight into Early Adopters and beyond, as Innovators handle risks of new innovations without requiring experience and typically understand and apply complex technical knowledge without prompting and move through the Innovation–Decision process without needing much prompting. Chapter 3 studies PEV financial incentive design and in turn pertains to Early Majority and beyond; as a generalization Early Adopters have a greater degree of upward social mobility than later adopters and prior research also points out that, for Early Adopters, financial incentives did not impact their adoption decision [24]. Finally, Chapter 4 provides insight into Late Majority and beyond with specific insights into used PEVs, which offers opportunity for those with broader socioeconomic status to consider PEVs with less financial risk and limited resources.

These paradigms are not only theoretical, as innovative technology adoption theory and "S curve" diffusion have been observed in practice with the adoption of many innovations to consumer goods. These useful theories provide an outline for diffusion that has proven accurate for many consumer product innovations when they are introduced to the market: technologies like electricity, radio, TV and cell phones have all entered the market and gained fairly rapid adoption [25].

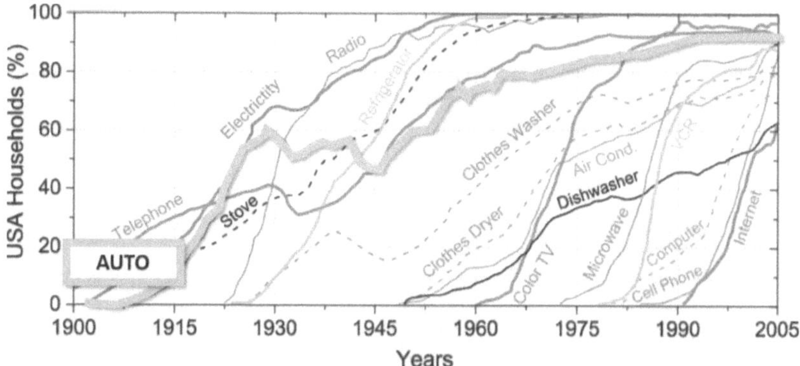

Figure 1-7: Historical Adoption of Innovative Technologies to US Households (%)
Figure from Carvalho et. al. [26] digitized from Nicholas Felton, New York Times [27] with automotive adoption highlighted by author

Automotive adoption has followed a similar S-curve adoption rate with some

variability: while adoption grew rapidly at first, growth rates faltered due to the market

circumstances of WWI and WWII, constrained purchasing power, manufacturing ability

and commodities shortages. After these macroeconomic factors passed, adoption

accelerated again, then slowed pace and took 45 years to gain ~10% of utilization. Now

US household utilization of automotive vehicles is ~90%, an expensive yet critical part of

everyday life for almost everyone in the US. With this has come the massive oil

consumption and environmental impact from having our personal transportation system

based on Internal Combustion Engine (ICE) technologies. Now that zero-emissions

alternatives, particularly PEVs are being developed and industrialized to a level not

previously seen, applying the theory of "S-curve" diffusion dictates that we are moving

through the "early adopter" beginning phases of innovative technology adoption and

these vehicles will soon be mainstream. As demonstrated by the adoption curve of

conventional ICE vehicles, however, adoption is subject to constraints over time, whether

it be market conditions or product and consumer limitations. When discussing PEV "S-curve" innovation diffusion it is critical that we start considering, evaluating, and preparing for the inevitable adoption plateaus of PEVs in the market so we can formulate the policy and strategic planning to push forward to the next phase and truly deliver mass-market zero-emissions transportation to consumers.

1.2 Consumer Acceptance of PEVs

Prior research has identified multiple barriers to achieving greater PEV adoption, including high purchase prices [28], [29] insufficient recharging infrastructure [30]–[32] and "range anxiety"—the fear that the driving range on a single charge will be insufficient to make it to a destination [16], [28], [30], [33], [34]. Researchers have also found that consumers often hold multiple misperceptions about PEVs—that they are less powerful than ICE vehicles, have worse environmental benefits, and are inconvenient to recharge—perceptions that are inconsistent with actual PEVs available on the market today [17], [35]. Finally, researchers have found that most consumers lack any basic knowledge of many aspects of PEVs, including their appearance, purchase price, acceleration performance, top speed, recharging time, driving range, recharging operation, electricity costs, and maintenance [4], [21], [36], [37].

Expanding on that point, knowledge has been shown in previous research to be an important component of automotive purchasing decisions with a significant impact on automotive purchases generally. As the established technology, the cost of ownership and driving experience of an ICE is well known, and comparisons on these items to PEVs have been suggested as an important factor affecting attitude and intentions, as well as something consumers struggle to assess correctly for fuel-efficient vehicles [17], [38], [39]. Research into PEV knowledge specifically has concluded that consumers are not very knowledgeable on many aspects of PEVs and there is much work needed to improve consumer knowledge these vehicles [4], [21], [36], [37]. Participants were not

knowledgeable on several specific aspects of PEVs, including appearance, price, top

speed, recharging time, driving range, recharging operation, electricity costs, and daily

maintenance [4], [37]. And at even a more basic level, 86% of households could not

differentiate between different PEV vehicles and technologies and the majority did not

understand the charging process [36]. These studies have found the lack of PEV

knowledge extends through many important customer-facing aspects of PEVs from price

and features of PEVs to general PEV technology operation.

1.3 Impact of Incentives on PEVs

Price associated with PEVs is another often cited issue for PEV adoption and high

purchase price is specifically cited as a factor constraining PEV adoption. In a study

analyzing consumer willingness to pay for PEV technology, researchers found that high

purchase price remains consumers' main concern about electric vehicles [28]. Another

study found that average US consumers are not willing to pay more for an PEV compared

to an ICE vehicle despite the technology's price premium [29]. Many of the studies on

consumer attitudes also include purchase price as another factor identified by consumers

as a reason not to consider an PEV [4], [16]. A few studies have captured the importance

of price sensitivity when it comes to PEVs and vehicle ownership that, "Vehicle

ownership costs are important in vehicle adoption choice for both private and business

purchases" [40], [41]. In general consumers do not always have accurate or adequate

information about possible fuel or maintenance cost savings (or other cost benefits),

therefore leading to flawed decision making [42]. As noted by Turrentine, "One effect of

this lack of knowledge and information is that when consumers buy a vehicle, they do not

have the basic building blocks of knowledge assumed by the model of economically rational decision-making, and they make large errors estimating gasoline costs and savings over time" [39]. This is particularly applicable to PEVs as there are fuel and maintenance savings over time along with other benefits of the technology.

A mechanism by which to address the high price of PEV technology and vehicles is incentives. In the US, the government offers a federal tax credit of $7,500 subject to individual tax situations and other criteria [43]. This legislation has recently evolved, still offering up to $7,500 tax credit with more qualifiers and thresholds for local production, battery and mineral content [44]. Certain states have additional incentive benefits including vehicle purchase incentives, reduced registration fees, reduced charging rates, charger incentives, emissions inspection exemption, HOV lane access, and designated/free parking. Vehicle purchase incentives typically take the form of rebates, subsidies, income tax credit, excise tax credit, or sales tax exemption [45]. Other researchers have grouped these as vehicle purchase incentives intended to reduce the purchase price of a PEV and the remainder as "reoccurring" or "ownership" incentives that apply throughout ownership of the PEV [46]. Each state in the US creates its own legislation related to PEV ownership, contributing to a high amount of difference in the structure, amount and execution of state PEV incentives, with some states even implementing PEV fees [47]. Between federal and state differentiation and variation, a convoluted system in the US has developed around PEV incentives as a mechanism to increase consumer adoption.

1.4 Used PEVs Market and Resale Value

So far adoption of PEVs has been explored through the lens of consumer behaviors as it related to new vehicle market and sales, however there is a secondary, used-car segment that plays an important role in the US automotive market. Average vehicle transaction prices have been trending upward in the US, with the average

transaction price in sales to consumers topping $35,655 in September 2020 and growing

month over month [48]. The secondary market has emerged as a practical means to

purchase affordable vehicles. The used-car market is also over double the size of the

new-car market in the US and has consistently delivered over twice the amount of cars

since 2012 [49], [50].Furthermore, replacement of older, used ICE vehicles with a used

PEV would deliver further environmental benefits with the removal of a less efficient

gasoline vehicle from the US fleet and subsequent adoption of a zero-emission option. As

the secondary market continues to develop and expand for PEVs, this will be a cost-

effective and further market opportunity that reaches additional consumers, many of

whom may not ever have considered PEV ownership a possibility.

To summarize, there is much to be gained from further research into PEV

adoption and the various consumer elements that contribute to future success. This book

covers distinct inputs, with insights gained into PEV consumer experience, knowledge,

incentivization, and resale value as well as spanning across adoption curve phases. A

better understanding of today's PEVs and consumers gives us the opportunity to offer

improved policies and strategies for the growth of this innovative technology moving

into the future.

Chapter 2: Impact of PEV Experience and Knowledge on Consumer Acceptance

"Electric vehicle adoption: can short experiences lead to big change?" in *Environmental Research Letters.*

Meeting the goals of the Paris Agreement will require massive decarbonization of the transportation sector—the largest contributor to anthropogenic greenhouse gas emissions in the US[1]. Plug-in electric vehicles (PEVs) offer a promising pathway to rapid decarbonization, provided they are charged on low-carbon energy sources. Despite a wide range of government incentives to increase PEV adoption, sales are still low compared to conventional Internal Combustion Engine (ICE) vehicles [2 – 4]. In 2018, US PEV sales comprised just 2.1% of total vehicle sales [51], and during this timeframe with the exception of Tesla the combined monthly sales of battery electric vehicles (BEVs) sold by all other automakers have been flat for the past five years (see Figure 2-1).

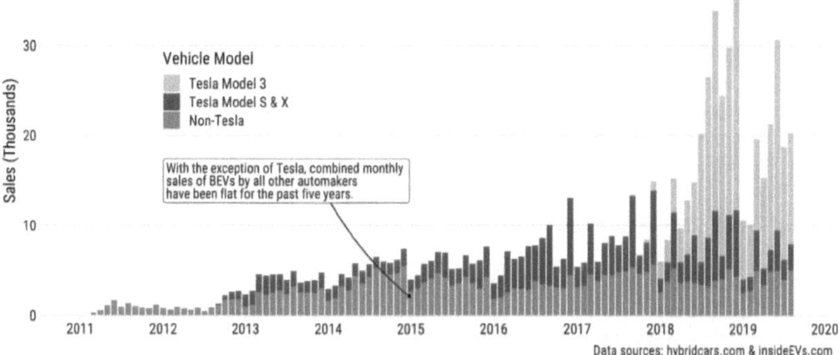

Figure 2-1: US Monthly Sales of BEVs.
Despite Tesla's market success, PEVs comprised just 2.1% of vehicle sales in 2018. The authors developed this figure using vehicle sales data from hybridcars.com and insideEVs.com.

Prior research has identified multiple barriers to achieving greater PEV adoption, including high purchase prices [6 – 7], insufficient recharging infrastructure [8 – 10], and "range anxiety"—the fear that the driving range on a single charge will be insufficient to make it to a destination [6], [8], [11 – 13]. Researchers have also found that consumers often hold multiple misperceptions about PEVs, including that they are less powerful than ICE vehicles, have worse environmental benefits, and are inconvenient to recharge—perceptions that are inconsistent with actual PEVs available on the market today [14 – 15]. Finally, researchers have found that most consumers lack basic knowledge about many aspects of PEVs, including their appearance, purchase price, acceleration performance, top speed, recharging time, driving range, recharging operation, electricity costs, and maintenance [4], [16 – 18].

One potential strategy to help alleviate some of these barriers and misperceptions is to increase consumers' direct experience with PEVs. Prior research has found that consumers who have had direct experience with PEVs were more comfortable with the technology, noticed more of the advantages that PEVs offer, and in general perceived PEVs more positively than those who lacked direct experience [19 – 22]. Experiments that measure differences in stated perceptions about PEVs before and after having a direct experience with a PEV have concluded that direct experience results in a more positive perception and opinion about PEVs and their performance [23 – 30]. Other studies found that participants with more PEV experience could readily recognize the environmental and economic benefits of PEVs, such as lower refueling costs, and were more capable of assessing whether their true driving range needs would be met by a PEV [19 – 21], [31]. Finally, studies also found that after directly experiencing a PEV, participants had more favorable opinions about PEVs and higher stated purchase intentions [25], [27], [29 – 30].

Nonetheless, these studies have all involved relatively long time frames—from days to months—during which participants experienced a PEV, which limited the feasible sample size (most studies have had less than 100 participants). The implications of these studies are also limited in their scalability; for example, it would be unreasonably costly for many thousands of customers to test drive a PEV for days or months in order to increase PEV adoption. Table 2-1 places this study in the context of this prior literature (a more detailed and comprehensive version is provided in Appendix Table A- 9).

Table 2-1: Summary of Studies on the Effect of Direct PEV Experience on PEV Perceptions.

Author	Year	Location	Year	Time Frame	Sample Size	Change in Perception
Gärling	2001	Sweden	1998 - 2000	3 months	42	No change
Carroll	2010	UK	2010	NA	69	+
Turrentine et. al.	2011	LA, NY, NJ	2009 - 2010	12 months	102	+
Burgess et. al.	2013	UK	2008 - 2012	6 – 12 months	55	+
Jensen et. al.	2013	Denmark	2012	3 months	369	+/-
Bühler et. al.	2014	Berlin	2009 - 2010	6 months	77	+
Franke	2014	Germany	2014	3 months	29	No Change
Wikström et al.	2014	Sweden	2011-2012	18 months	50	+
Skippon et. al.	2016	UK	2016	36 hours	393	+/-
Schmalfuß et. al.	2017	Germany	2017	24 hours	30	+
This study	**2019**	**Washington D.C.**	**2019**	**3-5 minutes**	**6,518**	**+**

In this study, we aim to assess the effect of *short* exposure times (e.g. minutes) on stated PEV purchase consideration. By limiting the exposure time to riding in a PEV for just three to five minutes, we were able to achieve a much larger sample size ($n = 6,518$) compared to similar prior studies. We find that the short experience of riding in a PEV on average had a significant, positive effect on participants' stated consideration ratings for adopting a PEV. Whereas longer duration experiences expose fewer participants to a wider variety of situations, our findings suggest that a single, shorter duration experience may be effective in increasing overall PEV adoption consideration across a larger population.

2.1 Methods

We collaborated with an industry partner, EZ-EV—a start-up subsidiary of Exelon Corp, one of the largest energy providers in the U.S.—to conduct a PEV ride along experience at the 2019 Washington DC Auto Show. The mission of EZ-EV is to simplify the process of consumer consideration and adoption of PEVs by providing educational and shopping tools about PEVs and hosting PEV events.

The PEV ride along experience was conducted inside the Walter Reed Convention Center from April 18-26, 2019 and was open to all attendees of the Auto Show, although individuals under 18 had to be accompanied by a parent or guardian. The PEV experience involved riding in a PEV with a professional driver around a short indoor course for approximately three to five minutes to experience some of the features of PEVs, including a 0-40 mph section to specifically highlight the acceleration performance and drivability of the vehicle. During the experience, the drivers answered questions specific to the vehicle and about PEVs in general. While they were not provided a script to follow, we know that some drivers provided basic information regarding the vehicle's drivability, the charging process, and available incentives based on feedback from participants that we interviewed after the experience. It is certainly possible that some of this information could have influenced the participants' survey responses, but isolating this effect is a limitation of the experiment. As a result, our results must be interpreted as the joint effect of riding in a PEV with an informative driver.

The available vehicles included three battery electric vehicles (BEVs)—the Audi e-tron, Hyundai Kona, and Nissan Leaf—which run entirely on electricity and can be recharged from the grid; one plug-in hybrid electric vehicle (PHEV)—the Toyota Prius Prime—which combines a conventional gasoline-powered engine with a battery that can be recharged from the grid; and one fuel cell electric vehicle (FCEV) —the Hyundai Nexo—which use fuel cells powered by hydrogen to produce electricity for the motor.

Data was collected at the driving experience booth through entry and exit surveys. All participants were required to take the entry survey and register for the event first, during which time they were given a unique ID code. The entry survey included three sections: 1) information about their current and future vehicle(s), 2) questions about their knowledge of PEVs, including the maximum available federal subsidy and vehicle refueling requirements, and 3) a Likert scale rating of two questions: a) whether they would "consider" a BEV and a PHEV for their next vehicle, and b) whether they would "recommend" either vehicle type to a friend. Respondents were not informed about whether their answers to the knowledge questions were correct or not. Table 2-2 summarizes the questions asked on the entry and exit surveys (a complete copy of the surveys can be found in the Appendix A, Section A.5).

Table 2-2: Summary of Entry and Exit Survey Questions.

	Entry survey	**Exit survey**

Survey-specific questions	• Demographics:
	  – Current vehicle (year, make, model, age)
	  – Time to next vehicle purchase
	  – Number of vehicles owned
	  – Home parking access
	  – Whether neighbor owns PEV
	• PEV knowledge questions:
	  – Vehicle types that can be fueled with gasoline
	  – Vehicle types that can be plugged in
	  – Maximum available federal subsidy
		• Which vehicle(s) rode in.
		• Brands considering for next purchase.

Questions asked in both surveys	• ID Code • Consideration and recommendation ratings for BEVs & PHEVs	

After taking the entry survey, participants were driven around the course in one of the vehicles by a professional driver. Participants were given the option to choose which vehicle to ride in, otherwise they were randomly assigned to one. After the driving experience was over, participants completed the exit survey, which captured the vehicle(s) the participants rode in as well as the same consideration and recommendation questions as shown in the entry survey.

The final dataset for our analysis was formed by using the unique respondent ID codes to match the entry and exit survey responses. To assess the impact of the PEV experience on the participants' consideration and recommendation ratings for BEVs and PHEVs, we first tabulated the ratings provided before and after the experience. Then, to examine the effect of the experience on the probability of participants choosing each rating level, we estimated an ordinal logistic regression (also known as the "proportional odds" or "log odds" model), which incorporates the inherent ordering of the consideration ratings that participants could choose, from "Definitely not" to "Definitely yes."

To explain the model, let Y be an ordinal outcome with J categories, corresponding to each rating level in the survey. If $P(Y \leq j)$ is the cumulative probability of Y being less than or equal to rating level $j = 1, \dots, J - 1$, then the *odds* of being less than or equal to rating level j is defined as

$$\frac{P(Y \leq j)}{P(Y > j)}, \quad \text{for } j = 1, \dots, J - 1 \tag{2.1}$$

The log odds, also known as the "logit," is then defined as

$$\log \frac{P(Y \leq j)}{P(Y > j)} = logit[P(Y \leq j)], \quad \text{for } j = 1, \dots, J - 1 \tag{2.2}$$

The ordinal logistic regression model defines a linear relationship between equation (2.2)
and a series of independent variables. The specific model we use is given by the
following equation:

$$logit[P(Y \leq j)] = \alpha_j - \beta x - \sum_i \gamma_i z_i - \sum_i \delta_i z_i x, \qquad (2.3)$$

where $j = 1, ..., J - 1$ and $i = 1, ..., M$ independent variables. The α_j coefficients in
equation (2.3) are intercepts that represent the dividing points between each level of the
ordered ratings *before* participants had the PEV experience (e.g. α_1 denotes the division
between "Definitely not" and "Probably not"). The β coefficient determines the
significance and magnitude of the main effect of interest: the before / after effect of
having the PEV experience, where x is a dummy variable for the time period (0 for
"before" and 1 for "after" the experience). Thus, if α_j determines the probability of
choosing each rating *before* the PEV experience, then $\alpha_j - \beta x$ determines the probability
of choosing each rating *after* the PEV experience. The γ_i coefficients reflect the effect of
other independent variables, z_i, on the before experience rating. These include the
experience, knowledge, and demographic variables listed in Table 2-3, such as correctly
answering a knowledge question or having parking access at home. Finally, the δ_i
coefficients reflect the interaction effect between the independent variables z_i and the
time period variable, x. These terms reflect how much the main before / after effect
(given by β) changes based on the independent variables z_i. For example, participants
that rode in a more premium vehicle, such as the Audi e-tron, might be expected to have

a larger *change* in their ratings than those that rode in less premium vehicles, all else being equal.

Since the outcome variables in this model (the log odds) is not immediately intuitive to interpret, we convert them into probabilities of choosing a rating by taking the inverse logit:

$$P(Y \leq j) = \frac{exp(\alpha_j - \beta x - \sum_i \gamma_i z_i - \sum_i \delta_i z_i x)}{1 + exp(\alpha_j - \beta x - \sum_i \gamma_i z_i - \sum_i \delta_i z_i x)} \tag{2.4}$$

We then use the values of each $P(Y \leq j)$ level to compute the individual probabilities of choosing each rating level. For example, the probability of choosing rating level 1 ("Definitely not") is $P(Y \leq 1)$, and the probability of choosing rating level 2 ("Probably not") is $P(Y \leq 2) - P(Y \leq 1)$, and so on, with the probability of choosing rating 5 ("Definitely yes") being equal to 1 minus the sum of the others.

We estimate a series of models to assess how different variables influence the probability of choosing a rating before and after the PEV experience. Each model is estimated using the polr() command from the MASS package in R [64]. The first model tests the main effect of interest: the time period before and after the experience. We also estimate models to control for each of the vehicle metrics we captured in the survey as well as the knowledge questions. The full set of variables are shown in Table 2-3. In the results section, we only report results for models that had non-negligible outcomes on the consideration ratings for BEVs. While we did also collect ratings on PHEVs, the results do not vary substantially compared to those from the BEV ratings. We include these results and other additional model results in Appendix A, Section A.3). The raw data and code to reproduce all results can be found at Github.

Effect	Variable	Description	Units / Values
PEV experience metrics	*timePeriod*	Rating given before or after PEV experience.	Before (0); After (1)
	etron, kona, leaf, nexo	Variable for each vehicle model rode in (base level is the Toyota Prius Prime).	For each vehicle model: Yes (1); No (0)
	PHEVpowertrain, FCEVpowertrain	Variable for powertrain of vehicle rode in (base level is BEV).	Yes (1); No (0)
	countCarsDriven	Variable for number of vehicles ridden in.	1 to 5
Respondent demographics	*homeParking*	Dedicated home parking spot (i.e. accessible for home charging).	Yes (1); No (0)
	neighborPEV	Whether respondent's neighbor owns a PEV.	Yes (1); No (0)
	multicar	Owns more than 1 vehicle.	Yes (1); No (0)
PEV knowledge	*bothFuels*	Correctly answered both refueling questions.	Yes (1); No (0)
	pluginFuel	Only correctly answered plug-in refueling question.	Yes (1); No (0)
	gasFuel	Only correctly answered gasoline refueling question.	Yes (1); No (0)
	subsidy	Correctly answered subsidy question.	Yes (1); No (0)

2.2 Results

2.2.1 Participant Sample and Consideration Ratings

Out of the 7,509 people that participated in the experience, 6,518 respondents completed both the entry and exist surveys—a completion rate of 86.8%. To keep the survey short and facilitate throughput, respondents were only asked about their current vehicle (age and make), when they plan to purchase a vehicle next, and which vehicles they rode in during the experience. The majority of participants had vehicles that were less than 10 years old, which aligns with the average light duty vehicle age in the US of 11.8 years [65]. The most common brands owned in the sample were Toyota, Honda, Ford, Chevy and Nissan, and most respondents stated not being in the market for a new vehicle. A total of 7,787 total rides were taken during the experience (some participants rode in more than one vehicle). Despite having five vehicles available, 90% of respondents rode in either the Audi e-tron or the Hyundai Kona as they were available every day of the event. The other cars were not always available due to various factors, such as driver availability or technical issues with the vehicle. We summarize this information about the sample in Table 2-4.

Table 2-4: Summary of Sample Demographics.

Current Car Age	NA	1 year or less	2-5 years	6-10 years	11-20 years	21+ years
Total Reponses:	366	1,132	2,694	1,805	1,329	183
Complete Reponses:	298	968	2,336	1,593	1,164	159
Current Vehicle Make	Toyota	Honda	Ford	Chevrolet	Nissan	Other
Total Reponses:	897	887	541	473	410	4,301
Complete Reponses:	779	776	465	414	355	3,729
Time to next purchase	Not in Market	0-6 Months	6 to 12 Months	12+ Months		
Total Reponses:	4,809	432	695	1,573		
Complete Reponses:	4,140	370	617	1,391		
Vehicle rides taken	e-tron	Kona	Leaf	Nexo	Prius Prime	Total
Complete Reponses:	3,117	3,925	597	135	13	7,787
Neighbor has an PEV	Yes	No	I'm not sure			
Total Reponses:	1,458	4,249	1,802			
Complete Reponses:	1,275	3,681	1,562			
Number of cars in Household	0	1	2	3	4	5
Total Reponses:	410	3,487	1,762	973	508	369
Complete Reponses:	332	3,010	1,535	858	449	334
Dedicated home parking	Yes	No				
Total Reponses:	4,420	3,089				
Complete Reponses:	3,872	2,646				
Number of Vehicles Ridden In	1	2	3	4	5	NA
Complete Responses:	5,284	1,100	92	3	3	36

While we asked respondents to rate both their consideration to purchase a PEV and whether they would recommend a PEV to a friend, we only present results for the *consideration* questions here because 1) the recommendation ratings were similar to the consideration ratings, and 2) our primary interest is consumer adoption of PEVs.

Furthermore, because results were similar for BEVs and PHEVs, we only show results

for BEV ratings (results for PHEVs can be seen in Appendix A, Section A.1 PHEV

Response Results). Figure 2-2 shows the change in BEV consideration ratings before and

after the PEV experience.

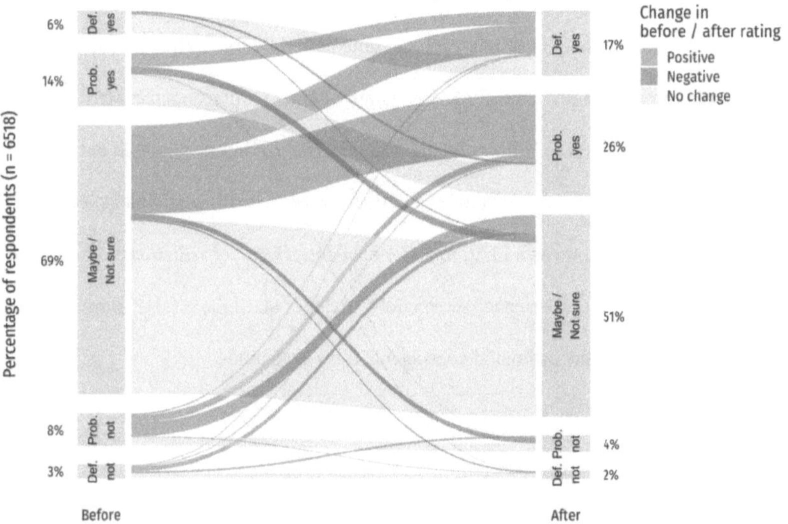

Figure 2-2: Change in BEV Consideration Rating Before and After Ride Along Experience.

While the majority of respondents chose the same rating before and after riding in a

PEV (gray ribbons), those who changed their rating were far more likely to choose a

more positive rating (blue ribbons) than more negative rating (red ribbons). Before the

PEV experience, nearly 70% of respondents chose "Maybe / not sure", 11% chose

negative ratings, and 20% chose positive ratings. In contrast, the post-PEV experience

responses show a shift away from the "Maybe / not sure" rating (down to 51%) coupled with a shift towards positive ratings (up to 43%) and a decline in negative responses

(down to 6%).

2.2.2 Modeling Consideration Rating Choices

We estimated six different ordinal logistic regression models to assess the impact

of the PEV ride along experience and other factors on the BEV rating choices. Model 1

includes only the main effect of interest—the time period (before / after the PEV

experience); models 2a and 2b include effects for whether participants correctly answered

the knowledge questions for PEV refueling (2a) and the maximum available federal PEV

purchase subsidy (2b); model 3 includes an effect for whether the participants stated

having a neighbor who owns a PEV; model 4 includes effects for which vehicle models

they rode in during the experience; and model 5 includes all of the effects from models

1–4. Table 2-5 shows the estimated coefficients from each model.

Table 2-5: Estimated Coefficients from Ordinal Logistic Regression Models of BEV Ratings

		Model #:	1	2a	2b	3	4	5
		Description:	Before/After PEV Experience	PEV Knowledge: Fueling	PEV Knowledge: Subsidy	PEV Neighbor	Car Model(s) Rode In	Full Model (models 1 – 4)
		N:	13,036	13,036	13,036	13,036	13,036	13,036
Main effects	β	timePeriod	1.033 (0.036) ***	1.087 (0.047) ***	1.073 (0.039) ***	1.051 (0.041) ***	0.951 (0.100) ***	1.014 (0.105) ***
	γ_i	pluginFuel		-0.034 (0.080)				-0.106 (0.080)
		gasFuel		0.115 (0.084)				0.032 (0.084)
		bothFuel		0.47 (0.071) ***				0.256 (0.074) ***
		subsidy			0.785 (0.074) ***			0.653 (0.079) ***
		neighborPEV				0.620 (0.064) ***		0.559 (0.064) ***
		etron					0.098 (0.065)	0.045 (0.065)
		kona					0.029 (0.067)	0.007 (0.067)
		leaf					0.219 (0.097) *	0.157 (0.097)
		nexo					-0.031 (0.184)	0.007 (0.185)
Interaction effects	δ_i	pluginFuel		-0.114 (0.109)				-0.106 (0.110)
		gasFuel		-0.028 (0.114)				-0.002 (0.114)
		bothFuel		-0.205 (0.095) *				-0.142 (0.101)
		subsidy			-0.253 (0.100) *			-0.208 (0.106)
		neighborPEV				-0.039 (0.087)		-0.016 (0.087)
		etron					0.093 (0.088)	0.114 (0.088)
		kona					0.081 (0.091)	0.095 (0.091)
		leaf					-0.151 (0.133)	-0.129 (0.133)
		nexo					0.168 (0.250)	0.155 (0.252)
Intercepts	α_j	definitelyNot	probablyNot	-3.232 (0.056) ***	-3.151 (0.060) ***	-3.146 (0.057) ***	-3.131 (0.057) ***	-3.149 (0.089) ***	-3.003 (0.091) ***
		probablyNot	maybeNotSure	-1.937 (0.033) ***	-1.860 (0.039) ***	-1.854 (0.034) ***	-1.835 (0.035) ***	-1.855 (0.076) ***	-1.711 (0.078) ***
		maybeNotSure	probablyYes	1.365 (0.029) ***	1.456 (0.036) ***	1.478 (0.031) ***	1.498 (0.032) ***	1.450 (0.075) ***	1.655 (0.078) ***
		probablyYes	definitelyYes	2.682 (0.037) ***	2.780 (0.043) ***	2.811 (0.039) ***	2.828 (0.040) ***	2.768 (0.078) ***	3.005 (0.082) ***

Significance codes: ***=0.001, **=0.01, *=0.05

29

Across all models, the main effect—the before / after effect of the PEV

experience—is large, statistically significant, and robust to the inclusion of other effects.

All other statistically significant effects in each model are smaller in magnitude than the

before / after effect. To make this effect easier to interpret, we use equation 2.5 to convert

the estimated coefficients in model 1 into probabilities of choosing each rating level

before and after the PEV experience. Figure 2-3 shows these probabilities, with error bars

computed using simulation (details are provided in Appendix A, Section A.2 Additional

Details on Model Estimation Methods).

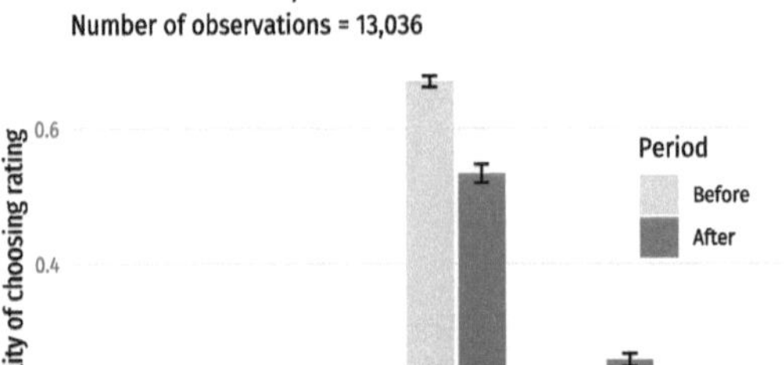

Figure 2-3: Predicted Probabilities of BEV Consideration Rating Choices Before and After PEV Experience.
The predicted ratings shift towards more positive ratings after the experience. Error bars
represent a 95% confidence interval reflecting uncertainty in the model parameters, computed
using simulation (see in Appendix A, Section A.2 Additional Details on Model Estimation
Methods).

In addition to the BEV rating questions, participants were asked three questions

assessing their knowledge about PEVs: two pertaining to the refueling requirements of

different vehicle types and one regarding the maximum federal subsidy available for

purchasing a PEV. The responses indicate that most participants were not knowledgeable about these aspects of PEVs. Only 30% of respondents correctly answered either the

plug-in or gasoline refueling question (with only 18% correctly answering both refueling

questions), and only 14% correctly answered the subsidy question, with 80% stating they

were not sure. Figure 2-4 summarizes the responses to these questions.

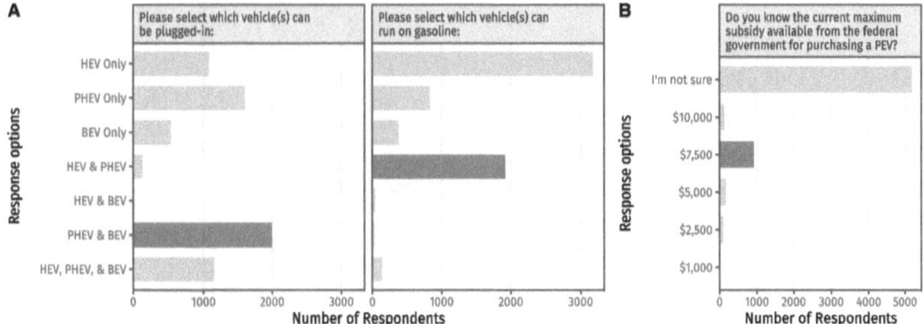

Figure 2-4: Knowledge Question Results (Correct Response Highlighted in Green).
Only 30% of participants correctly answered either the plug-in or gasoline refueling questions,
with just 18% correctly answering both (A). Only 14% of participants were able to correctly
answer the question about the federal PEV subsidy, with 80% stating they were not sure (B).

Although the majority of respondents did not correctly answer the knowledge

questions, those that did chose higher BEV consideration ratings both before and after the

PEV experience. Those who knew the maximum federal PEV subsidy available in

particular had higher ratings than those who correctly answered the refueling questions,

though this may be expected as respondents who knew the subsidy may already be

considering purchasing a PEV. Nonetheless, these results suggest that even those with

more knowledge about PEVs were still on average chose more positive ratings after the

PEV experience. Figure 2-5 shows the results of models 2a and 2b converted to

probabilities of rating choices.

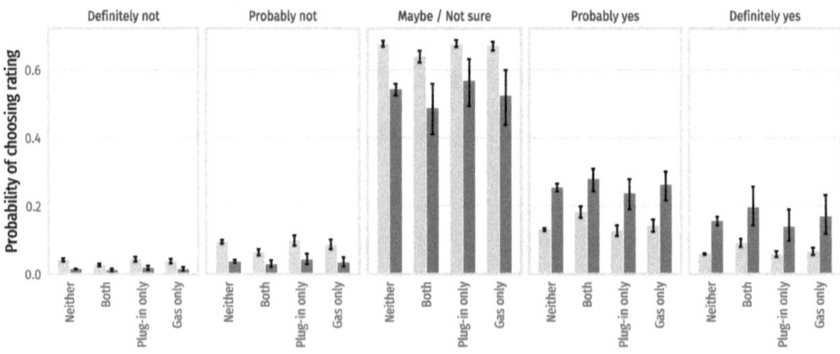

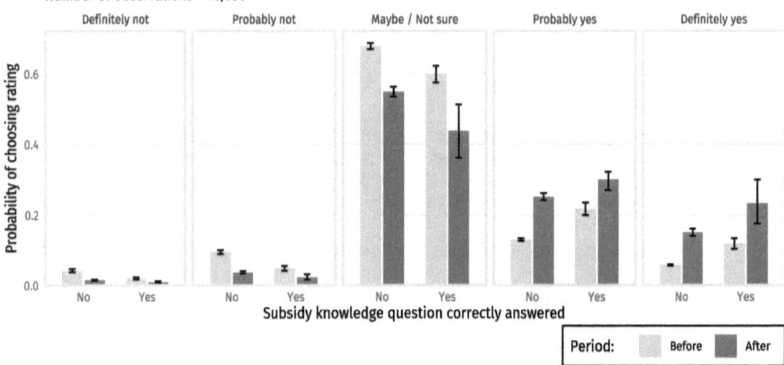

Figure 2-5: Predicted Probabilities of BEV Consideration Rating Choices Before and After PEV Experience.
Respondents with greater knowledge about PEV refueling (A) and subsidies (B) had higher predicted ratings both before and after the experience. Error bars represent a 95% confidence interval reflecting uncertainty in the model parameters, computed using simulation (see in Appendix A, Section A.2 Additional Details on Model Estimation Methods).

Results of model 3 suggest that respondents that stated they had a neighbor that owns a PEV chose higher BEV consideration ratings both before and after the PEV experience (this effect is similar in size to that of those who correctly answered the subsidy knowledge question). This corresponds with prior research on the "neighbor effect," which suggests that people are more likely to consider adopting a PEV if their

neighbors have also adopted a PEV [34 - 35]. Figure 2-6 shows these results.

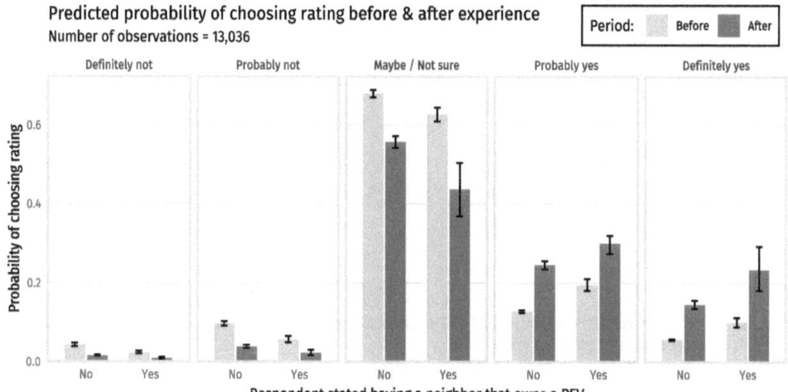

Figure 2-6: Predicted Probabilities of BEV Consideration Rating Choices Controlling for Neighbors with PEV.
Those who indicated their neighbors own a PEV showed greater levels of consideration before and after the PEV experience, with the PEV experience still providing an increase in consideration. Error bars represent a 95% confidence interval reflecting uncertainty in the model parameters, computed using simulation (see in Appendix A, Section A.2 Additional Details on Model Estimation Methods).

Finally, results from model 4 suggest that the particular vehicle model that

participants rode in did not have a substantial effect on their rating choices, as shown in

the overlapping error bars in Figure 2-7. This is important as it suggests that a short ride

in a much more affordable PEV, such as the Nissan Leaf, could potentially be just as

significant as riding in a luxury PEV in terms of influencing peoples' consideration about

adopting a PEV.

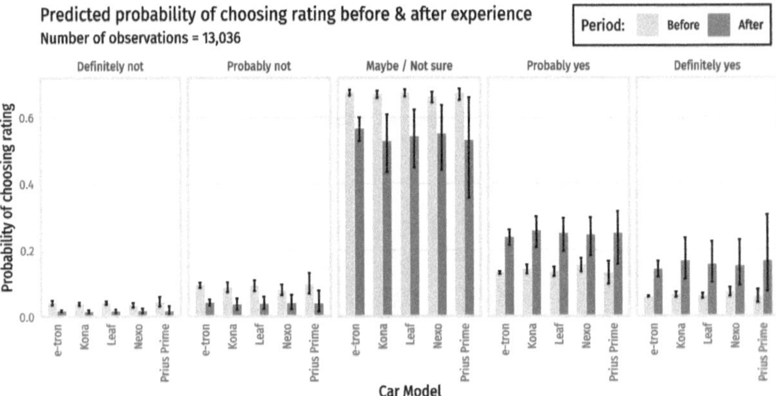

Figure 2-7: Predicted Probabilities of BEV Consideration Rating Choices Controlling for Vehicle Model Rode In during Experience.
Although we expected to see more positive ratings for riding in a luxury vehicle model, such as the Audi e-tron, we did not find evidence of this. Instead, it appears that the post-experience positive rating was approximately the same regardless of the vehicle ridden in. The error on the Prius Prime is substantially larger due to the fact that few respondents rode in it. Error bars represent a 95% confidence interval reflecting uncertainty in the model parameters, computed using simulation (see in Appendix A, Section A.2 Additional Details on Model Estimation Methods).

Model 5 provides a comprehensive model including all variables from models 1 – 4. While we did estimate additional models to control for other factors, such as having dedicated at-home parking and how many cars the participant owned, none of these models had statistically significantly different results. Results from these additional models can be found in Appendix A, Section A.3 Additional Models.

2.3 Discussion and Conclusions

In this study, we conducted a large-scale experiment investigating how a short, direct exposure with a PEV could impact participants' stated consideration of adopting the technology. Overall, results indicated that the experience had a positive impact on their stated consideration of and recommendation for PEVs. Our results show a 118% increase in the number of participants that stated they would "probably" or "definitely" consider purchasing a BEV for their next vehicle—from 1,268 participants before the experience to 2,762 after. These results agree with previous studies that used substantially longer exposure times in their experiments [23 – 30]. For example, in Carroll et al. [58], 72% of participants stated they would use a PEV as their regular car after the test drive experience compared with just 47% before the test drive [58], and in Turrentine et al. [61] 67% of respondents changed their opinion about PEVs after the end of their leases, with 71% of respondents stating they would be more likely to purchase a PEV after the experience than before. Part of why this short exposure experience may have had such an impact on participants' consideration of PEVs is that it familiarized participants with an otherwise foreign vehicle technology. Rogers (2003) suggests that knowledge and experience acquisition of a new technology leads to lower perceived risk from it and more favorable intentions towards it [9]. The short, scalable experience provided in our experiment may have helped address some of the known knowledge and risks issues associated with PEV adoption [16].

An important observation of these results is that even a short exposure time with a PEV may lead to a significant increase in positive PEV perceptions—a result that has important practical implications for increasing PEV adoption. The type of short, experiential event that we conducted is well within the scope of the types of events vehicle manufacturers and dealers currently conduct with conventional vehicles in the form of promotional events. This structure could be utilized by automakers and dealers to promote PEVs in a cost- and time-effective manner to increase consumer acceptance of PEVs on a much larger scale. Fleet operators such as taxi and ride hailing companies could also potentially impact a much larger portion of the general public by operating

PEV fleets and exposing riders to the technology. Policymakers could support PEV usage in fleets and other forms of short-term direct exposure events.

In terms of the general public's knowledge about PEVs, our results corroborated prior research on large gaps in PEV knowledge and misperceptions that exist [4], [21], [36], [37]. Research in this area also found that greater PEV knowledge was associated with more positive perceptions of PEVs, lower perceived risks of PEVs, positive evaluation of PEV attributes, and a higher intent to pay a premium for renewable fuel— all of which are associated with a great overall intention to adopt a PEV [4], [18], [22], [38 – 39]. Our study suggests similar results. Participants that correctly answered the knowledge questions on the survey—in particular, the federal subsidy question—had higher PEV consideration ratings both before and after the experience, suggesting that knowledge about the technology may be important for shifting public option and willingness to consider adopting a PEV. More research in this area is needed to determine effective means for conveying this knowledge to consumers and what kinds of knowledge might have the biggest impact.

Although this experiment was successful in capturing a substantially larger sample size than prior research on direct experience with PEVs, the experiment design is potentially susceptible to response biases that can be present in survey-based studies. In particular, because all respondents were shown the same rating questions on both the before and after surveys, there is a chance that some of the shift to more positive ratings after the experience is due to response bias. We acknowledge this potential as a limitation of the study, and thus the main before / after experience effect should be interpreted with this potential bias in mind. Nonetheless, other aspects of the study suggest that response bias alone would not likely explain the results. For example, verbal responses from participants during short interviews conducted immediately after the PEV experience were overwhelmingly positive towards PEVs and frequently involved statements of surprise about PEVs, in particular about the high acceleration performance of the PEV. In addition, some respondents stated more *negative* ratings after the experience than they did before. Finally, some of the results should be robust to response bias. For example, the association between having greater knowledge about PEVs and stating higher consideration ratings could still be concluded as the *difference* between the more and less knowledgeable participants should not be affected by response bias. A more robust research design would have had a random sample of participants only respond to a post-survey. This would have allowed us more ability to analyze different control groups and quantify any potential biases in the response from seeing the same questions in both surveys.

Some limitations in the data collection process include the lack of more detailed demographic data (a trade-off made between the time required to complete the surveys and participant throughput) and the self-reported nature of the demographic data collected. For instance, 20% of the complete sample reported that their neighbor owns a PEV, which is likely higher than reality given the historically low PEV sales rate. Given that few participants correctly answered the knowledge questions about PEVs, we suspect that many who responded "yes" likely had neighbors with vehicles that they mistook for a PEV, such as a hybrid vehicle. Nonetheless, while it is not possible to determine the accuracy of their responses, it may still be valuable to gauge the difference in the perceptions of those who *believe* they encounter PEVs on a daily basis with those that do not.

Finally, it is important to address potential limitations in the generalizability and longevity of the effects noted in this study. Participants in this study were individuals who opted to attend the DC Auto Show and participate in the EV ride along experience, presenting a limited and self-selecting sample. There is potential for self-selection bias in our sample; participants were already opting to engage in an automotive event and elected to participant in our ride and drive, possibly creating a sample not applicable beyond these circumstances. However, the DC Auto Show is open to the general public with thousands of attendees and attracts a wide range of individuals. In addition, other ride and drive events for specific automakers were also present at the event, including Jeep, Jaguar, and Land Rover. Based on responses to our knowledge questions and results from the study, there was a similar lack of knowledge and experience with PEVs that is consistent with the general population. In addition, qualitative verbal responses from participants after the experience were overwhelmingly consistent with a positive overall experience. Unfortunately, we are unable to assess the longevity of the effects captured in this study as we did not obtain identifiable information about the respondents necessary to contact them at a later time. This is an important consideration for future related work and could be easily integrated into future research in this area.

Electrifying the vehicle fleet to meet carbon goals will require PEV adoption are far greater rates than have historically occurred. This will require solutions that go beyond attracting early adopters and scale to attract the general public. Our study shows that having a direct experience—even for just a few minutes—with a PEV could be

important for changing public opinion about PEVs on a mass scale.

Chapter 3: Consumer Preferences of PEV Financial Incentives

"Not all subsidies are equal: measuring preferences for electric vehicle financial

incentives" in *Environmental Research Letters.*

Plug-in electric vehicles (PEVs) are an important pathway for decarbonizing the transportation sector, yet sales in the US are still relatively low. Federal and state incentives, such as purchase subsidies, have been shown to have a measurable effect on increasing PEV adoption [70], but how these incentives are designed can affect both their value to customers as well as how equitably they are distributed [71]. In this study, we aimed to measure how US vehicle buyers value different features of PEV financial incentives to identify incentives that are more attractive and more likely to be distributed across a more diverse group of consumers.

Most studies on the impacts of PEV incentives find that financial incentives lead to increased PEV adoption. In a 2017 review of studies on PEV incentives, 32 of 35 studies concluded that PEV subsidies have a positive effect on PEV sales [70], mirroring earlier research on the effectiveness of financial incentives in increasing hybrid sales [72]–[74]. In addition, evidence suggests that financial incentives are becoming even more important over time as more mainstream car buyers start adopting PEVs [75]. One reason is that average US consumers in general do not appear to be as willing to pay a premium for PEVs compared to wealthier early adopters; evidence suggests that PEVs would indeed need to have considerably lower prices than those of internal combustion engine vehicles in order to be equally desirable for mainstream buyers, all else being equal [29].

A particularly important finding from this body of research is that not all incentives are equally effective in increasing new technology adoption. One of the earliest studies on adoption rates of hybrid electric vehicles found that providing consumers with sales tax exemptions had a more than ten-fold increase in sales compared with providing income tax credits [73]. Literature reviews by Hardman et al. and Deshazo highlight multiple studies that find a similar preference for tax exemptions over tax credits as well as a preference for incentives that are applied at the point of sale over those applied post-sale [70], [76]. These results are consistent with the well-known phenomenon of "present bias," a cognitive bias in which money is valued more in the present than in the future [77].

Incentive design also impacts how equitably incentives are distributed. In a review of the distributional effects of US clean energy tax credits, Borenstein and Davis found that incentive programs aimed at incentivizing PEV purchases were the most extreme in terms of the incentive distribution, with the top income quintile receiving approximately 90% of all credits [78]. This outcome is unsurprising considering how the federal subsidy tax credit is structured. In general, time-delayed incentives like tax credits are skewed towards higher-income buyers who can afford the full up-front PEV purchase price. Furthermore, not all households are eligible to receive the full PEV incentive amount ($7,500 for most full electric vehicles) as it depends on tax liability. Specifically, households with lower tax liabilities due to income or the availability of other credits, such as the child tax credit, may not receive the full credit amount compared to high-income earners with larger tax liabilities. In a study on Atlanta, GA, researchers estimate that only 23% to 45% of the city population could qualify for the full PEV tax credit, and these percentages are even lower for minority populations [71].

Finally, while the vast majority of PEV tax credits have gone to the wealthiest buyers [79], studies suggest that a more equitable allocation to buyers with lower and middle incomes may have been more effective at increasing overall PEV adoption. A survey by Hardman et al. found that financial incentives were not an important decision factor for purchasers of high-end PEVs but were significantly important for purchasers of lower-end PEVs [24]. Likewise, in a quasi-experimental analysis of subsidies in California, Muehlegger and Rapson found relatively large demand elasticities for PEVs among low-income buyers, with an estimated 32-34% increase in demand for every 10% reduction in purchase price via a subsidy [80]. The combined outcomes of these prior studies suggest that improvements could be made to incentive designs to make them more attractive and equitable.

3.1 Methods

We aimed to measure how US vehicle buyers value different features of PEV financial incentives to identify incentive designs that are both more valuable to consumers and more likely to be distributed across a more diverse group of consumers. To do so, we designed and fielded a nationwide choice-based conjoint survey online in August and September of 2021. In conjoint surveys, respondents are asked to choose their most preferred option from a set of alternatives in a series of consecutive choice questions. Each alternative in each choice question is comprised of a list of attributes (e.g., "price") with different levels (e.g., different dollar amounts). We use a randomized survey design, meaning that the attribute levels shown in each choice set were randomly chosen from the full set of combinations of all the levels for each attribute. While this approach requires a larger sample size to obtain precise parameter estimates compared to alternatives that attempt to maximize information, such as "D-optimal" designs, it allows for greater flexibility in the types of models that can be estimated [81]. The choice data obtained can then be used to estimate choice models to quantify the relative value respondents hold for each attribute shown.

By using a controlled experiment, we are able to disaggregate preferences for different incentive features and explore heterogeneity in those preferences among different sub-populations, which can be difficult (if not impossible) using historical incentive data as they have limited variation amongst incentive features. Furthermore, our survey results reveal preferences for the general car buying population as opposed to wealthy early adopters. Given the diversity of car buyers in our sample, we report effects for different subgroups in the sample, which is important for any revisions to current subsidy policy as PEVs are gradually adopted by more diverse populations.

We restricted the survey to financial incentives because these incentives have been found to be one of the more effective types [70] and because including other non-financial incentives would require a substantially larger sample size to identify the preferences of sub-groups within the sample, which is important for understanding the equity implications of the study. Based on preliminary piloting and reviews of prior literature, the following incentives features were included in the choice questions: 1) Type—a tax credit, tax deduction, sales tax exemption, or a direct rebate; 2) Amount—the total dollar amount of the incentive, ranging from $1,000 to $8,000 in increments of $500, which reflects the range of historically available incentive amounts (for the sales tax type, the amount is an exemption of 50%, 75%, or 100% of the sales tax, computed as the participants self-reported budget multiplied by 7.5%); 3) Timing—the time when the incentive will be received (immediately versus several levels of delay, ranging from weeks to the next tax filing period); and 4) Source—the government, dealership, or automaker. Each participant answered 10 consecutive conjoint questions. We also asked additional questions regarding respondent demographics as well as their knowledge about and experience with PEVs. Respondents were screened such that they were 18 years old or older, reside in the U.S., and were in the market for a vehicle. The full text of the survey is available in Appendix B, Section B.3.

To ensure that participants understood the choice task, respondents were first shown a practice question where the rebate option was the logically dominant choice. This question was used to screen out respondents as choosing anything other than the dominant choice suggests they were likely either not paying close attention or did not understand the choice task. Figure 3-1 shows example choice questions.

A.

Great let's practice!

If you could only choose ONE incentive, which would you prefer? Note: You casnnot "stack" multiple incentives.	Sales Tax Exemption	Tax Credit	Tax Deduction	Rebate from *Government*
	Amount: $1,000	Amount: $1,000	Amount: $1,000	Amount: $8,000
View incentive descriptions in new tab	Time Frame: Time of Sale	Time Frame: At Tax Filing (Approx. April 2022)	Time Frame: At Tax Filing (approx. April 2022)	Time Frame: Time of Sale From: Government

B.

(2 of 10) Which incentive option would you choose?	Sales Tax Exemption	Tax Credit	Tax Deduction	Rebate from *Government*
	Amount: $1,900	Amount: $1,000	Amount: $2,000	Amount: $1,500
View incentive descriptions in new tab	Time Frame: Time of Sale	Time Frame: At Tax Filing (approx. April 2022)	Time Frame: At Tax Filing (approx. April 2022)	Time Frame: 6 weeks after purchase From: Government

Figure 3-1: Example Conjoint Questions.
Figure A is the practice question shown to all respondents, and Figure B is an example randomized question.

The conjoint choice questions were designed and randomized using the *cbcTools* R package [82], and the survey was implemented on formr.org, an open-source online platform that uses the R programming language to define survey questions [83]. An initial pilot survey was fielded on Amazon Mechanical Turk (N = 216 participants) for basic testing purposes. The final survey was fielded using an online panel via Dynata, a market research firm. We applied a stratified sampling approach to match the income distribution of US car buyers for the first 2,000 respondents, and we collected an additional 500 respondents targeting those with household incomes below $50,000 as this group was under-sampled in the original run.

We accounted for cost of living (COL) differences across the national sample by adjusting the *amount* variable shown in the survey by a COL adjustment scalar prior to estimating all models. The COL scalar for each respondent was obtained by matching each respondent's self-reported zip code to its associated Core-Based Statistical Area (CBSA) using data from the US Department of Housing and Urban Development (HUD) [84] and then matching each CBSA to a COL adjustment factor from the Real Personal Income by State and Metropolitan Area dataset provided by the US Bureau of Economic Analysis [85]. This resulted in the *amount* value being scaled down by a factor less than 1 in locations where the COL is higher than the national average (since the value of a dollar buys less in places with a higher COL) and scaled up by a factor greater than 1 where the COL is lower than the national average. This COL adjustment had little effect on the model results (see Table B-4 in Appendix B for the un-scaled model results). After removing 338 respondents who did not have a zip code that matched with the COL adjustment data, our final sample was 2,170 respondents. Table 3-1 summarizes

demographic statistics of the final sample including COL adjusted results.

Table 3-1: Summary of Sample Demographics

Summary Statistics	N = 2,170
Age	
Min	19
Max	92
Mean	56
(NA)	6
Gender identity	
Male	1,211 (56%)
Female	942 (43%)
Other	11 (0.5%)
Prefer not to say	2 (<0.1%)
(NA)	4
Timeframe for purchase	
1 year	914 (42%)
0-3 months	786 (36%)
No timeline	470 (22%)
Shopping for new or used	
New	1,284 (59%)
Used / both / not sure	886 (41%)
Income	
> Median	1,404 (65%)
< Median	750 (35%)
Prefer not to say	12 (0.6%)
(NA)	4 (0.15%)

Using the choice data, consumer choice can be modeled using a random utility

framework, which assumes that individual consumer i makes choices among alternatives

j that maximize an underlying random utility model, u_{ij}, which can be parameterized as a

function of an alternative's observed attributes, v_{ij}, and a random variable representing

the portion of utility that is unobservable to the modeler, ε_{ij}, such that $u_{ij} = v_{ij} + \varepsilon_{ij}$.

For this study, the utility model for alternative j for individual i can be expressed as

follows:

$$u_{ij} = \boldsymbol{\beta}'\mathbf{x}_j - \lambda a_j + \varepsilon_{ij},$$

$$(3.1)$$

where λ is the coefficient for the incentive amount, a_j, and $\boldsymbol{\beta}$ is a vector of coefficients for all other attributes, $\mathbf{x}_j$. To make the results more easily interpretable, we specify the utility model in the "willingness-to-pay" (WTP) space [86], [87] such that estimated model coefficients represent the marginal WTP (or valuation in the context of this study) for marginal changes in each attribute:

$$u_{ij} = \lambda\left(\boldsymbol{\omega}'\mathbf{x}_j - a_j\right) + \varepsilon_{ij},$$

$$(3.2)$$

where $\boldsymbol{\omega}$ is the WTP coefficients for all non-price attributes, $\mathbf{x}_j$, and λ is now a scale parameter. Using the WTP space for the utility models has several conveniences. Since WTP coefficients have units of dollars, they can be immediately interpreted and understood independent of other parameters. In addition, since WTPs are independent of error scaling, they can be directly compared across different models estimated on different subsets of the data. In contrast, preference space coefficients represent marginal utility, which must be interpreted relative to other parameters and cannot be directly compared across models due to potential scaling differences. For the specific context of this study, the general WTP space utility model takes the following form:

$$u_{ij} = \lambda\left(\boldsymbol{\omega}'\mathbf{x}_j^{\text{type}} + \boldsymbol{\tau}'\mathbf{x}_j^{\text{timing}} + \boldsymbol{\eta}'\mathbf{x}_j^{\text{source}} + \boldsymbol{\omega}^{\text{tt}\prime}\mathbf{x}_j^{\text{type}}\mathbf{x}_j^{\text{timing}} + \boldsymbol{\omega}^{\text{ts}\prime}\mathbf{x}_j^{\text{type}}\mathbf{x}_j^{\text{source}} - a_j\right)$$
$$+ \varepsilon_{ij},$$

$$(3.3)$$

where ω, τ, and η are vectors of WTP parameters for each incentive *type*, *timing*, and *source*, respectively, and each of the x_j terms are the dummy-coded variables for these respective attributes. The incentives amount is given by a_j. Interactions terms for *type*timing* and *type*source* are included since the timing and source values vary depending on the type (e.g., the sales tax exemption is always at the time of sale). The *rebate* type was set as the reference level for the dummy-coded incentive types. For *timing*, the reference level was *time of sale* for the *tax credit* and *rebate* types (the timing didn't vary for the *sales tax* and *tax deduction* types). Finally, for *source*, the reference level was *government* for the *rebate* type (the source didn't vary for all other incentive types).

We assess consumer valuation for different financial incentive features by estimating multinomial logit (MNL) models on the full sample and subgroups within the sample as well as a mixed logit (MXL) model on the full sample via maximum likelihood estimation, a common and well-established estimation approach for discrete outcome utility models [88], [89]. The multinomial logit models assume fixed preference parameters across the survey population whereas in the mixed logit model preference heterogeneity is modeled according to parametric assumptions about the population preference parameters. One of the convenient features of the logit model is that by assuming the error term of the utility model follows a Gumbel extreme value distribution[1]

---

[1] Gumbel and independent normal distribution error assumptions yield practically identical results empirically, and Gumbel yields a more convenient expression for choice probabilities.

the probability that a consumer i will choose option j from the choice set $\mathcal{I}_c$ follows a convenient closed form expression [89]:

$$P_{ij} = \frac{exp(v_j)}{\sum_{k \in \mathcal{I}_c} exp(v_j)}, \qquad \forall c \in \{1,2,3 \dots C\}, \quad j \in \mathcal{I}_c,$$

$$(3.4)$$

where c indexes a set of C choice sets. For mixed logit, probabilities are approximated using random draws from parameter distribution via maximum simulated likelihood [89]. Panel effects from repeated choices by each individual are accounted for in the calculation of the log-likelihood function for MXL models (see equation 6.2 in Train, 2009 [89]), and standard errors are clustered at the individual level. All models were estimated using the *logitr* R package [90].

Finally, in our results we present some WTP values as the sum of multiple different WTP coefficients. For example, the WTP for a 6-week-delayed rebate from a dealership is the sum of the coefficients for *type_rebate*, the interaction parameter for *type_rebate*timing_6_weeks*, and the interaction parameter for *type_rebate*source_dealer*. For these calculations, 95% confidence intervals are computing by taking multivariable normal draws of the model parameters using the mean estimates and the variance-covariance matrix of the model, summing the appropriate draws to compute the desired value, and observing the 2.5% and 97.5% percentiles of those draws [91]. Code to reproduce analysis and results can be found on GitHub.

3.2 Results

We present the results from several models. Model 1 is a MNL of the full sample, model 2 is a MXL of the full sample, and models 3 – 5 are MNL models on subgroups based on demographic information, including high- vs. low-income buyers (3), new vs. used vehicle buyers (4), and high vs. low budgets (5). All models are estimated in the "Willingness to Pay" (WTP) space such that coefficients reflect preference values in dollars [86], [87]. Table 3-2 shows the estimated coefficients from each model, which are in units of thousands of dollars. Since each respondent answered 10 choice questions, the final dataset includes 21,700 choice observations from sets of four incentive types: sales tax exemption, tax credit, tax deduction, and rebate.

Results from all models suggest that car buyers value financial incentives significantly differently depending on how they are implemented. Coefficients from model 1 imply that participants overwhelmingly prefer immediate rebates, on average valuing them by $580, $1,450, and $2,630 more than sales tax exemptions, tax credits, or tax deductions, respectively (see Figure 3-2). The mixed logit model (model 2) suggests similar results, valuing immediate rebates by $650, $1,580, and $3,580 more than sales tax exemptions, tax credits, or tax deductions, respectively. The results that a sales tax exemption is valued at several hundred dollars less than an immediate rebate is particularly interesting as they both occur at the time of sale, suggesting there could be an intrinsic difference in value for receiving money compared to avoiding a fee. In addition, unlike an immediate rebate, the dollar savings from a sales tax exemption can depend on the purchase price.

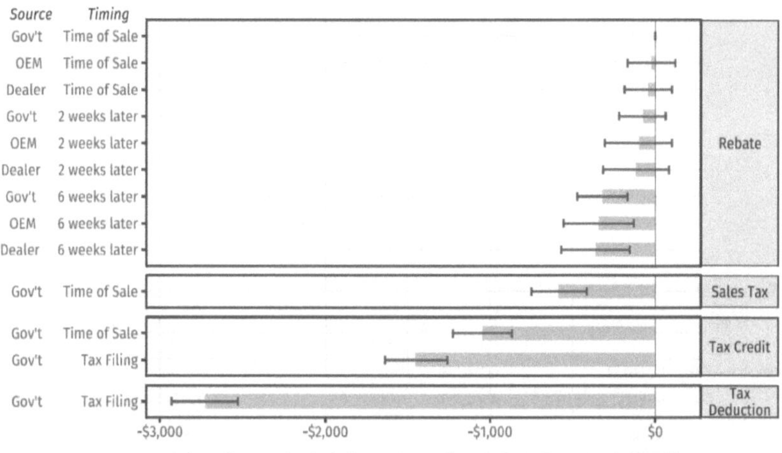

Value of Incentive Relative to Immediate Rebate from Gov't ($USD)

Figure 3-2: Value of Different Incentive Designs Relative to an Immediate Government Rebate at the Time of Sale.
Bars represent the mean WTP coefficients from model 1, and error bars reflect a 95% confidence interval computed via simulation as described in the Methods section [91].

Table 3-2: Summary of Estimated Model Coefficients.

Model:	(1) Multinomial Logit	(2) Mixed Logit (mean)	(st. dev)	(3a) Above Median Income	(3b) Below Median Income	(4a) New Car Buyers	(4b) Used Car Buyers	(5a) Budget >$30k	(5b) Budget <$30k
Respondents:	N = 2,170	N = 2,170		N = 1,404	N = 750	N = 1,284	N = 886	N = 1,125	N = 1,045
Scale parameter*	0.519 *** (0.0)	0.711 *** (0.0)	---	0.562 *** (0.0)	0.460 *** (0.0)	0.525 *** (0.0)	0.512 *** (0.0)	0.545 *** (0.0)	0.488 *** (0.0)
Sales Tax	-0.582 *** (0.1)	-0.648 *** (0.1)	2.235 *** (0.1)	-0.499 *** (0.1)	-0.742 *** (0.2)	-0.586 *** (0.1)	-0.563 *** (0.1)	-0.389 *** (0.1)	-0.858 *** (0.1)
Tax Credit	-1.449 *** (0.1)	-1.583 *** (0.1)	2.508 *** (0.1)	-0.982 *** (0.1)	-2.438 *** (0.2)	-1.143 *** (0.1)	-1.896 *** (0.2)	-0.973 *** (0.1)	-1.963 *** (0.1)
Tax Deduction	-2.727 *** (0.1)	-3.576 *** (0.1)	-2.800 *** (0.1)	-2.436 *** (0.1)	-3.261 *** (0.2)	-2.726 *** (0.1)	-2.718 *** (0.2)	-2.443 *** (0.1)	-3.048 *** (0.2)
Tax Credit: Immediate	0.403 *** (0.1)	0.297 ** (0.1)	-0.365 . (0.2)	0.239 ** (0.1)	0.789 *** (0.1)	0.140 (0.1)	0.798 *** (0.1)	0.198 * (0.1)	0.624 *** (0.1)
Rebate: 2-week delay	-0.075 (0.1)	-0.026 (0.1)	0.344 (0.2)	-0.016 (0.1)	-0.190 (0.1)	-0.081 (0.1)	-0.063 (0.1)	-0.000 (0.1)	-0.156 (0.1)
Rebate: 6-week delay	-0.318 *** (0.1)	-0.254 *** (0.1)	0.492 ** (0.2)	-0.231 * (0.1)	-0.510 *** (0.1)	-0.352 *** (0.1)	-0.273 * (0.1)	-0.143 (0.1)	-0.515 *** (0.1)
Rebate: Source OEM	-0.022 (0.1)	-0.148 . (0.1)	-0.574 *** (0.2)	-0.011 (0.1)	-0.014 (0.1)	-0.050 (0.1)	0.025 (0.1)	0.008 (0.1)	-0.057 (0.1)
Rebate: Source Dealer	-0.042 (0.1)	-0.102 (0.1)	-0.323 * (0.1)	0.012 (0.1)	-0.124 (0.1)	-0.112 (0.1)	0.057 (0.1)	-0.012 (0.1)	-0.079 (0.1)

Signif. Codes: '***' = 0.001, '**' = 0.01, '*' = 0.05, '.' = 0.1, ' ' = 1

*The incentive amount in all models was scaled to account for cost-of-living differences

The current federal tax credit was consistently one of the least-valued incentives, with an average devaluation of $1,450 (model 1) and $1,580 (model 2) compared to an immediate rebate. Even after adjusting the tax credit timing to be assignable at the point of sale, it is still over $1,000 less valuable to consumers compared to an immediate rebate. While this may seem surprising, it is important to again emphasize that many households may not be able to claim the full tax credit as it depends on their tax liability. In contrast, the direct rebate is immediately delivered, reducing the purchase price, taxes, and fees at the point of sale, and the amount provided is independent of any individual's tax circumstances.

Our results also indicate that the timing of when an incentive is received matters, with delays resulting is significant devaluations. For example, model 1 suggests that delaying a rebate by 2-weeks or 6-weeks lowers the incentive value by approximately $40 or $320, respectively. Using these results, we can compute the implied discount rate of the current maximum federal tax credit ($7,500) for different incentive types. Using coefficients from model 1 suggests that car buyers discount time-delayed incentives at rather large rates: 30%, 46%, and 54% for 2-week, 6-week, and 6-month delays, respectively. These rates are consistent with depreciation rates of passenger vehicles; for example, gasoline-powered vehicles typically lose as much as 60% of their initial value within their first five years of use, which is equivalent to a 20% discount rate [92]. These discount rates suggest that delaying incentives can significantly reduce their value to consumers.

Given that the vast majority of historical PEV subsidies have been allocated to wealthier car buyers [78], [79], making PEV incentives more equitably accessible is critically important for the design of future incentive policies. We compared the

preferences of respondents from households with annual incomes above and below the

median US income (model 3), which is $67,521 according to the 2020 census [93], those

who stated they were shopping exclusively for new vehicles versus those that were also

considering used vehicles (model 4), and those with larger versus smaller stated budgets

(model 5). Results show that an immediate rebate is even more highly valued for lower-

income buyers, used vehicle buyers, and those with smaller budgets. Specifically,

respondents with annual household incomes below the median income (N = 750) valued

a tax credit by $2,440 less than an immediate rebate whereas those with annual household

incomes above the median (N = 1,404) valued a tax credit by $1,000 less than an

immediate rebate (see Figure 3-3). Below-median income households also devalued the

tax deduction and sales tax exemption incentives at greater levels than higher-income

households. Similarly, used vehicle buyers (N = 886) valued a tax credit at $1,900 less

than an immediate rebate compared to only $1,140 for new car buyers (N = 1,284), and

respondents with a budget of less than $30,000 (N = 1,045) valued the tax credit at

$1,960 less than an immediate rebate compared to only $970 for those with higher

budgets (N = 1,125). Preferences for other incentive features were similar for each of

these groups.

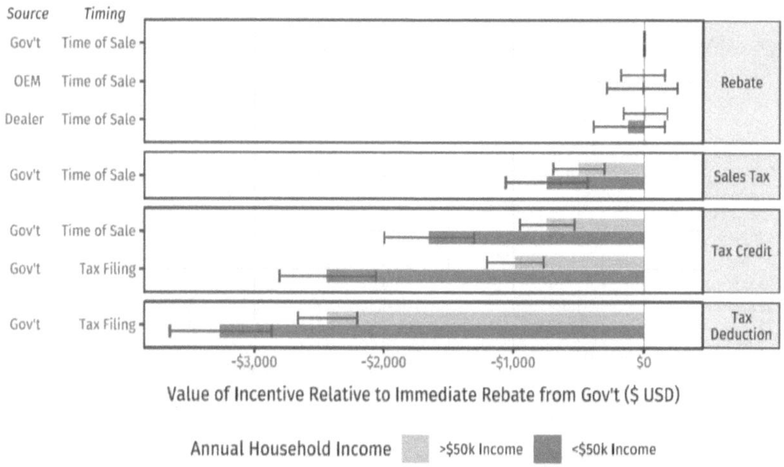

Figure 3-3: Value of Different Incentive Designs Relative to an Immediate Government Rebate at the Time of Sale for Above- and Below-Median Income Households. Bars represent the mean WTP coefficients from model 3, and error bars reflect a 95% confidence interval computed via simulation as described in the Methods section [91].

In addition, participants were asked whether they knew the current maximum federal tax credit, whether or not they would ever consider purchasing a BEV or PHEV, and whether or not their neighbors own a PEV (response summaries are shown in Table 3-3). Results from models comparing groups according to their responses to these questions (Table B-6 in Appendix B) suggest that the tax credit design is less attractive to buyers who are already less likely to purchase a PEV. Likewise, those who are already considering a PEV, know more about the current incentives, and have neighbors that own a PEV value those incentives more. All of these groups strongly preferred an immediate rebate over the current tax credit system. These results also suggest that improving education and awareness about PEVs and available incentives may be important for increasing the value of future incentives to customers as those who were more knowledgeable about PEVs valued the incentives more. Results of models estimated comparing other demographic groups largely had insignificant statistical differences

between groups, often due to small sample sizes in one or more group. These include

housing ownership, access to home parking, ethnicity, education, and work.

Table 3-3: Response Summary for PEV-related Questions.

Do you know the current maximum available federal subsidy?		Would you ever consider purchasing a PHEV?	
Not sure	1,685 (67%)	Definitely Yes	372 (15%)
$10,000	89 (3.6%)	Probably Yes	580 (23%)
$7,500	380 (15%)	Maybe / Not sure	820 (33%)
$5,000	206 (8.2%)	Probably Not	428 (17%)
$2,500	111 (4.4%)	Definitely Not	305 (12%)
$1,000	36 (1.4%)	(NA)	3
(NA)	1		
Do any of your neighbors own a PEV?		**Would you ever consider purchasing a BEV?**	
Yes	433 (17%)	Definitely Yes	334 (13%)
No	1,382 (55%)	Probably Yes	407 (16%)
Not sure	690 (28%)	Maybe / Not sure	678 (27%)
(NA)	3	Probably Not	559 (22%)
		Definitely Not	527 (21%)
		(NA)	3

Across every model we estimated, immediate rebates were valued significantly

more than tax credits. This suggests that the federal government could have achieved the

same value to PEV buyers with less taxpayer dollars if the federal PEV subsidy had been

implemented as an immediate rebate. To estimate this potential savings, we estimated the

total amount of federal tax credits available to all eligible PEVs sold between 2010 and

2019 (approximately 1.4 million PEVs) [94], [95]. Using PEV sales data at the make-

model level, we assumed that BEVs received the full tax credit amount ($7,500) and

computed the amount for PHEVs based on the battery capacity, accounting for the

subsidy phase out when specific models that reached the 200,000 sales limit [96]. This

results in an estimated $8.65 billion in tax credits. If this subsidy were instead

implemented as an immediate rebate, we estimate the federal government could have

saved approximately $2.07 billion (24%), or $1,440 per PEV on average (see Figure 3-4).

This implies that an immediate rebate program would still deliver greater value to

customers at the same cost to the government so long as the potential additional

administrative burden of implementing the program remains less than $1,440 per PEV.

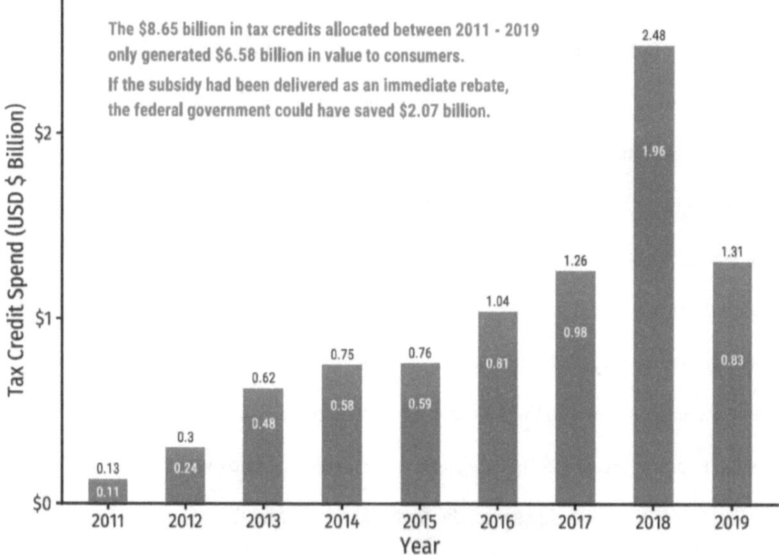

Figure 3-4: Estimated Annual Federal PEV Subsidy Allocation Between 2011 and 2019.
The red portion indicates the estimated amount the federal government could have saved if the
tax credit were delivered as an immediate rebate rather than a tax credit. Estimates are based on
applying the federal tax credit policy to every eligible PEV sold using sales data from
hybridcars.com and insideEVs.com.

3.3 Discussion

In this study, we aimed to understand how consumers value different features of PEV incentives to inform the development of a more effective and equitable incentive design. Based on the results of a nation-wide conjoint survey, we find that both the incentive *type* (tax credit, tax deduction, sales tax exemption, or direct rebate) and *timing* (at the point of sale or some period after purchase) significantly impacted its value while the *source* (government, dealer, or OEM) had little to no impact. Respondents valued more immediate incentives over time-delayed incentives, with discount rates ranging from 30% to 53% for time-delayed incentives. Immediate rebates delivered at the time of sale are the most-valued incentive design across all subgroups in our sample. Relative to the current federal tax credit incentive, immediate rebates are valued by as much a $1,450 more on average. These findings are consistent with prior research that suggests consumers prefer incentives that are applied at the point of sale over those applied post-sale [70], [76]. This also aligns with the financial concept of Time Value of Money where money in the present is valued more than an equal amount in the future due to inflation, investment potential, compounding interest etc. and the psychological factors that make it harder for humans to equate present vs. future values. All these factors combines to emphasize the value of providing financial incentives for PEVs immediately as rebates.

Research has also shown that the current tax credit strongly favors households with higher incomes and fewer children and is thus not equitably accessible [71]. Our results complement these findings and suggest that the valuation for immediate rebates is nearly twice as large for lower-income households compared to higher-income households and approximately 50% larger for used vehicle buyers compared to new vehicle buyers. This suggests that implementing the PEV subsidy as a direct rebate and extending it to used PEVs could be an effective strategy to encourage a more equitable adoption of PEVs.

Our results also suggest that improving education and awareness about PEVs may also increase the incentive value to potential PEV buyers; approximately 85% of participants who took our survey could not correctly identify the currently maximum federal subsidy amount for PEVs, and respondents that lacked this knowledge or were not considering purchasing a PEV valued the tax credit less. This is consistent with other research that suggests direct experience with PEVs can increase consumers' willingness to consider purchasing them [97]. Combining an education and awareness campaign along with changes to incentive policy that aligns with consumer preferences could result in a more effective and equitable incentive by a larger and more diverse population than today's tax credit.

Unfortunately, the structure of today's PEV subsidy—a tax credit delivered when filing taxes—was consistently one of the least-valued incentive types, with only a tax deduction being valued less. This finding is not necessarily surprising considering the inconveniences and hurdles associated with the tax credit design. First, if the taxes a PEV buyer owes are less than the maximum $7,500 credit, the buyer will not receive the excess amount. Furthermore, since the credit is delayed until the annual tax filling process, the PEV buyer must finance the full PEV purchase price along with any state sales taxes and fees evaluated at that price. These two factors can significantly reduce the total amount of incentive available to consumers.

Nonetheless, maximizing value to the customer may not be a prioritized objective when designing a PEV incentives; indeed, ease of implementation can often outweigh other factors. For example, including a new tax credit into the already well-established tax filling procedure is arguably a simpler system to adopt (and pass legislation for) than many other alternative designs.

Furthermore, the current implementation of the tax credit does not necessarily require the government to send funds to buyers but rather reduces the total tax revenue raised. As a result, overhead costs for implementing the tax credit are potentially lower than alternative designs.

Based on the consumer preferences from our experiment, we estimate that by delivering the federal PEV subsidy as a tax credit, the cumulative amount of subsidy available to prior PEV purchases was devalued by approximately 24% compared to if it were delivered as an immediate rebate, resulting in a loss in value to customers (or a potential savings to the government) of $2.07 billion. While an immediate rebate may incur higher administrative costs to deliver compared to a tax credit, it would result in a more equitably distributed subsidy and would deliver a greater value to customers, and it would still be at least as cost effective as the current tax credit design so long as additional administrative burdens are no greater than $1,440 per PEV.

While results from this research show some very interesting results in terms of consumers' valuation of PEV financial incentives, it is important to interpret these along with potential limitations. First, the research questions around PEV financial incentives requires a more complex conjoint design and some participants may not have understood the choice task. We addressed this as thoroughly as possible throughout the design and analysis phases, offering education around every attribute in the conjoint, reference materials available throughout conjoint questions and a test question to filter out those not understanding or paying attention. But there still could be an element of lack of familiarity or understanding in the responses. Also, this was a completely hypothetical scenario and while participants might respond during the survey, their actions in real life might be different. It wasn't possible to translate the questions from this survey into an actionable check for participants given the nature of the vehicle transaction and incentives in question. An additional follow-up with respondents to assess if a purchase has taken place and details would have added an ability to confirm responses against actual behaviors.

3.4 Conclusions

Despite their rapid growth in countries with more aggressive policies [98], PEVs comprise just 2-3% of the new vehicle market share in the US[95]. If the US is to catch up with other countries, PEV incentives will play an important role, especially in expanding PEV adoption to more diverse populations beyond wealthy early adopters. How these incentives are designed can affect both their effectiveness and accessibility to diverse populations.

In this study, we aimed to understand how consumers value different features of PEV incentives to inform the development of a more effective and equitable incentive design. Based on the results of a nation-wide conjoint survey, we found that an immediate rebate delivered at the time of sale is the most-valued incentive design across all subgroups in our sample. Relative to the current federal tax credit incentive, immediate rebates are valued by as much a $1,450 more on average, and this valuation is nearly twice as large for lower-income households compared to higher-income households. We also found that used vehicle buyers value an immediate rebate more than new vehicle buyers. Implementing a direct rebate for both new and used PEVs could be an effective strategy to encourage a more equitable adoption of PEVs.

We also estimate that by delivering the federal PEV subsidy as a tax credit, the cumulative amount of subsidy available to prior PEV purchases was devalued by approximately 24% compared to if it were delivered as an immediate rebate, resulting in a loss of $2.07 billion in value to customers or savings to the government ($1,440 per PEV sold on average). While an immediate rebate may incur higher administrative costs to deliver compared to a tax credit, it would result in a more equitably distributed subsidy and would deliver a greater value to customers. Finally, our results suggest that improving education and awareness about PEVs may also increase the incentive value to potential PEV buyers.

Ethical Statement

All participants in this research were 18 years or older and consented to participate in this study. This research was determined to be research that is exempt from IRB review under DHHS regulatory category 3 by the George Washington University IRB (#NCR213531).

Chapter 4: Investigation of PEV Resale Market & Depreciation with Policy Implications

J.P. "Battery-Powered Bargains? Assessing Electric Vehicle Resale Value in the United States"

Plug-in electric vehicles (PEVs) are a critical technology for decarbonizing the US transportation sector—now the nation's largest contributor to greenhouse gas emissions [1]. But their success as a substitute for gasoline-powered conventional vehicles (CVs) will depend on whether consumers are willing to purchase them. While higher up front purchase prices and limited driving ranges are frequently cited as barriers to adoption [16], [29], researchers have found that uncertainty in their resale value (or "resale anxiety" [99], [100]) remains an important consideration for consumers purchasing a technology with uncertain durability [101]. PEV affordability in the resale market will also play a crucial role in expanding PEV adoption beyond wealthier households, which currently comprise the vast majority of PEV owners [102]. As the used PEV market expands, improving our understanding of value retention among used PEVs is critically important for both consumers and policy makers looking to incentivize PEV adoption in the resale market.

Vehicle resale value is affected by a variety of factors, such as the vehicle make, model, year, mileage, condition, and trim, as well as features related to the market, such as the location and overall supply of used vehicles. For PEVs, resale value can be affected by additional factors, such as the all-electric driving range and battery condition [103], as well as PEV-specific policies, such as tax incentives in the new PEV market [104]. Industry and academic research estimating PEV value retention rates has concluded that while hybrid vehicles (HEVs) have depreciated at similar rates as CVs, plug-in hybrid vehicles (PHEVs) and battery electric vehicles (BEVs) have depreciated faster [105]–[108]. Table 4-1 summarizes prior research on PEV residual value.

Table 4-1: Summary of Studies Quantifying PEV Resale Value

Study	Model Years	MSRP Data	Resale Value Data	Resolution	Sample Size	Main Results
This study	2012-2018	EPA; carsheet.io	marketcheck	Daily listings	9,015,324	BEVs and PHEVs depreciate quicker than CV/HEV but is improving with more recent model years and higher ranges.
Rush et al. (2022) [109]	2012-2019	Edmunds	Edmunds TMV	Monthly time series	582,000*	CVs and HEVs consistent 3-yr retention; PHEVs and BEVs initially lower but increasing in retained value
Burnham et al. (2021) [110]	2013-2019	EPA	Edmunds TMV	1 TMV snapshot (July 2020)	686*	BEVs and PHEVs depreciate more quickly than HEVs and CVs
Hamza et al. (2020) [111]	2014-2019	KBB	KBB	Snapshot (2019)	72*	PHEVs and CVs hold value similarly; BEVs 11% lower retention over 5 years
Guo et al. (2019) [103]	2010-2016	Wards	Edmunds TMV	Snapshot (Q4 2016)	1,400*	PEV retention lower than gasolines equivalents. Tesla major exception with highest retained value over time.
Schoettle et al. (2018) [112]	2011-2015	EPA	KBB	Snapshot (Jan. 2018)	200*	PHEVs retained resale value equally as well as CVs (i.e., 0% average difference), and BEVs improved to an average of -5.7% difference in resale value compared to CVs
Tal et al. (2017) [113]	2011-2015	New car buyers survey / OEM website	Self-reported used car buyers survey	Snapshot (2016)	160*	PEVs models held 34% (2011 Nissan Leaf) to 80% (2014 Toyota Prius plug-in) of value in 2015 compared to MSRP.
Zhou et al. (2016) [114]	Unknown	NADA guides	NADA guides	Unknown	Unknown	Comparing the adjusted retention rates of PHEVs and BEVs with those of CVs indicates 2-3 year retention rate is lower for PEVs.

*Sample sizes estimated based on descriptions of data in papers.
Abbreviations:
EPA = Environmental Protection Agency (fueleconomy.gov)
TMV = True Market Value (private party data)
KBB = Kelly Blue Book (private party data)
NADA = National Automobile Dealers Association

Although more rapid PEV depreciation is a consistent finding, this outcome is not necessarily the same for every vehicle. Tesla BEVs, for example, have been found to have among the highest value retention rates of any vehicle [103], [109], in part due to their desirable features, such as high driving ranges and well-established private charging infrastructure, but also due to supply constraints they have historically faced in the new market, leading to higher used vehicle price [115]. Furthermore, many of the studies concluding that PEVs rapidly depreciate are based on data from earlier PEV models between 2011 and 2016, a time when most BEVs had relatively low driving ranges and were first generation vehicles [103], [112], [113]. More recent studies are finding some evidence that newer model PEVs are holding their value better [109].

In this study, we aim to improve upon these prior studies using listing prices from a large, nationally representative dataset of used cars listed online between 2016 and 2020 in the United States. The listing data are licensed from marketcheck.com, a market research firm that collects vehicle listing data from dealership websites. Whereas prior studies have used smaller samples of pre-processed data, such as the Edmunds True Market Value [103], [109], [110], or snapshots of available listed vehicles [111], [112], [113], the data we use contains the raw, daily listing prices from 66,641 dealerships. The detailed data enables the ability to quantify changes in value retention due to vehicle features (e.g., mileage), environment features (e.g., number of days the vehicle was listed), policy features (e.g., available subsidies), and changes over time (e.g., newer versus older model years), enabling greater insights into PEV retention rates compared to prior studies. To control for other features not included in the listings data, we added electric driving ranges and MSRPs (used to calculate retention rates) from carsheet.io [116]. Vehicle operating costs (in cents per mile) were computed using vehicle efficiencies from fueleconomy.gov [117] and monthly gasoline prices [118] and annual average electricity prices [119] in different states from the US Energy Information Administration (EIA), using utilization factors (0 to 1) from fueleconomy.gov to compute the electric and gas portions of operating costs for PHEVs. We also include data on the federal and state subsidies available for new PEVs at the time vehicles were listed. More detailed descriptions of these calculations are included in the methods.

For our primary analyses, we censor the data to only include vehicles with ages between 1 and 8 years old as few BEV listings were present in the dataset outside of this period. While the data go out to March 2022, we only use listings up until the end of 2019 to avoid the pricing disruptions experienced due to supply shortages during and after the COVID-19 pandemic, which began in early 2020 (Figure 4-5 shows prices through March 2022 for different vehicle types for comparison). We also limit the data to vehicle models that comprised at least 1% of the listings within each powertrain, ensuring a representative sample of the majority of common vehicle models on the market while removing exotic models. We also focus on cars since few BEV pickups or SUVs were listed in the time period captured in the dataset. Table 4-2 summarizes the dataset by powertrain, with Tesla and non-Tesla BEVs separated given Tesla's historically higher retention rates and significant size in the US market. Appendix Table C-1 summarizes each car model included in our analyses.

Table 4-2: Summary of Used Vehicle Listings.

Key Sample Stats	Conventional N = 8,395,000	Hybrid N = 464,560	PHEV N = 58,915	BEV Tesla N = 22,518	BEV Other N = 74,331
Model Year					
2012	764,383	61,590	6,823	390	3,943
	(9.1%)	(13%)	(12%)	(1.7%)	(5.3%)
2013	1,188,624	93,803	14,342	2,797	11,245
	(14%)	(20%)	(24%)	(12%)	(15%)
2014	1,466,956	86,350	14,043	2,969	13,665
	(17%)	(19%)	(24%)	(13%)	(18%)
2015	1,942,194	102,049	8,217	6,645	25,007
	(23%)	(22%)	(14%)	(30%)	(34%)
2016	1,598,340	54,178	4,474	6,608	15,193
	(19%)	(12%)	(7.6%)	(29%)	(20%)
2017	1,007,898	45,596	9,905	1,232	4,075
	(12%)	(9.8%)	(17%)	(5.5%)	(5.5%)
2018	426,605	20,994	1,111	1,877	1,203
	(5.1%)	(4.5%)	(1.9%)	(8.3%)	(1.6%)
Mileage (1,000)					
Mean	42	45	42	34	24
Median	37	39	36	31	23
SD	26	28	25	19	14
Age (years)					
Mean	3.40	3.70	3.86	3.82	3.66
Median	3.28	3.58	3.65	3.80	3.61
SD	1.49	1.51	1.40	1.28	1.13
Listing price (Used $USD)					
Mean	16,192	16,954	16,684	51,314	14,109
Median	14,957	16,429	16,364	49,006	12,428
SD	6,188	5,040	4,347	13,111	6,068
MSRP (New $USD)					
Mean	28,104	30,193	37,765	88,157	36,606
Median	27,308	28,923	36,645	89,320	34,409
SD	7,819	4,806	3,284	8,825	5,134
Electric Range (miles)					
Mean	--	--	35	238	87
Median	--	--	38	246	82
SD	--	--	14	47	27
Minimum	--	--	11	139	58
Maximum	--	--	53	335	238

4.1 Methods

4.1.1 Data and Code Availability

All of the code used to process the data, estimate models, and produce all analyses

are publically available on GitHub. The vehicle listings data that support the findings of

this study are available from marketcheck.com, but restrictions apply to the availability of

these data, which were used under license for the current study and so are not publicly available. A sample of the data is included in the GitHub repository. The relevant variables in the full original database can be provided on an individual bases for review purposes only to reproduce the study results by contacting the lead contact. All other data used in the study are posted in the repository.

4.1.2 Data Preparation

We use vehicle listing data licensed from marketcheck.com, a market research firm that collects vehicle listing data from individual dealership websites daily (the same database was also used in a recent publication in *Joule* on PEV mileage[120]). Since the PEV market has a limited number of SUV models, we limit our analyses to only car models. We also limit our dataset to vehicle ages between 1 and 8 as fewer BEV listings are available outside of this range (fewer vehicles are listed used within 1 years of being new, and few used BEVs are older than 8 years old as of December 2019). Although we have access to data up to February 2022, we censored the data for our primary analyses to end in December 2019 due to the significant market impact of the COVID19 pandemic on vehicle pricing. In addition, we also only include vehicle models that comprised at least 1% of the listings within each powertrain as a practical compromise between including a representative sample of vehicles while remaining computationally reasonable as the majority of the listings are comprised of a smaller number of models and a large number of models have very few listings (e.g. exotic cars). The final dataset includes 9,015,324 unique used car listings from 66,641 dealerships. Appendix Table C-1 summarizes the dataset.

In addition, other vehicle specifications were obtained and joined onto the listings data from a variety of data sources. Vehicle MSRPs are from carsheet.io [116] and BEV and PHEV ranges as well as all vehicle efficiencies (miles per gallon for gasoline-powered vehicles and kWh per 100 miles for electricity-powered vehicles) are primarily from fueleconomy.gov [117] with a small amount of missing values added from carsheet.io. Monthly gasoline prices[118] and annual average electricity prices[119] in different states are from the US Energy Information Administration (EIA) which are used to calculate cents by miles operating costs which were also integrated. Pricing was also adjusted to account for inflation using Consumer Price Index[121]. Finally, we also include data on the federal and state subsidies available for new PEVs at the time used vehicles were listed. Data on these incentives between 2016 and 2018 were taken from Wee et al.[122] and manually updated through 2020 primarily from the Alternative Fuels Data Center.

4.1.3 Modeling Vehicle Retention Rate

To establish a comparable metric across different vehicles from different price ranges, we compute a retention rate for every vehicle model, calculated as the listing price divided by the original manufacturer's suggested retail price (MSRP) based on the vehicle model year (both listing prices and MSRPs were inflation-adjusted to constant 2019 dollars). While some studies subtract off available federal and state subsidies from the MSRP before computing the retention rate [103], [109], we choose not to make this adjustment for several reasons. First, subtracting off the available subsidies from the MSRP is a non-linear transformation of the retention rate that affects both the intercept and slope of the resulting depreciation curve (both intercepts and slopes will increase). We believe this is inconsistent with the effect that an initial subsidy in the new market should have, which is to lower the initial price but not necessarily affect the rate by which the value drops from that price. Subtracting off the subsidy in computing the retention rate could lead to a misleading conclusion with respect to annual depreciation rates, implying that the retention rates of subsidized vehicles (BEVs and PHEVs) are falling at faster annual rates, albeit starting from higher initial retention rates. Finally, the subsidy available at the time a used vehicle was initially purchased is arguably less relevant to its resale value compared to the subsidy for the equivalent *new* model at the time the used vehicle is listed. For example, the maximum price a used BEV could be listed for should be the price for the equivalent model in the new market minus available subsidies, otherwise the buyer would simply buy the new version and obtain the subsidy. Thus, in our models we include the total subsidy (state plus federal) available for the equivalent new model at the time the used vehicle is listed as a separate covariate to assess any potential effect of new PEV subsidies on listing prices in the resale market.

We develop an exponential decay model to estimate the potential effects of various factors, such as the make, model, mileage, model year, and powertrain, and available new market subsidies, on the retention rate of different vehicles. The basic exponential model follows that of other related studies [109], [110]:

$$r = \alpha e^{(\beta x)}$$

(4.1)

where r is the retention rate, x is a matrix of model covariates, β is a vector of coefficients for the covariates, and α represents the initial retention rate upon immediate sale in the new market. Taking the log of both sides of this equation yields a model that can be estimated using linear regression:

$$\log r = \alpha + \beta x$$

(4.2)

Interpreting the estimated $\hat{\boldsymbol{\beta}}$ coefficients in a meaningful way (e.g. the change in r based on a change in any one covariate in $\mathbf{x}$) requires the transformation $e^{\hat{\beta}} - 1$. We present the estimated (non-transformed) coefficients in Table C-2 and Table C-3 and the transformed coefficients in Table 4-3, Table 4-4, and Table C-4.

4.2 Results

Our primary variable of interest is the value retention rate, which is computed as the listing price divided by the new MSRP[2]. Figure 4-1 compares the retention rates of CVs (gray bands) with those of other powertrains (green bands), including HEVs, PHEVs, and BEVs, where the median (solid lines) and interquartile ranges (bands) of retention rates were computed for all listings in each month of age. In general, CVs and HEVs follow similar retention rate patterns over time while PHEVs retain slightly more value at first then drop off steeper as compared to CVs. BEVs in general depreciate faster than CVs, with the exception of Tesla BEVs (blue bands) which actual retain higher values in the first few years of age relative to CVs before falling to similar values with CVs. This aligns with prior research in this area which finds Tesla BEVs follow a different depreciation trend than other BEVs [103].

---

[2] In contrast to other studies, available subsidies were not subtracted off the MSRPs when computing retention rates (see the experimental procedures for a detailed explanation of this decision).

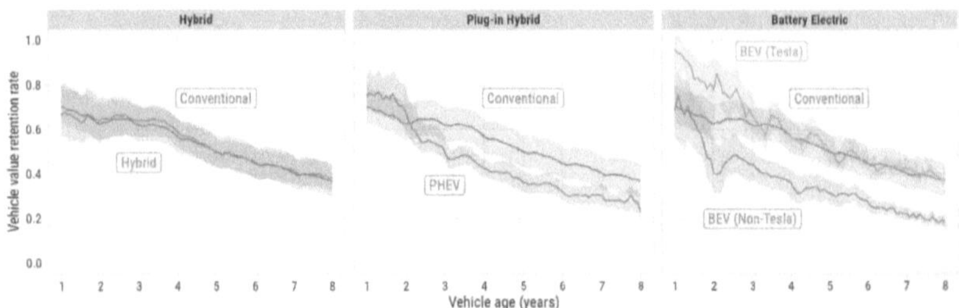

Figure 4-1: Comparison of the Median and Interquartile Ranges of Computed Retention Rates by Powertrain and Age.
The solid line shows the median retention rate and the bands reflect the 25th and 75th percentiles. The same curve for CVs (in grey) is shown for comparison in each sub-figure.

To quantify the differences in retention rates by powertrain, we estimate an exponential decay model of the log of retention rate on age interacted with powertrain type. Since the estimated coefficients (presented in Appendix C: Table C-2) are less intuitive to interpret, we present a transformation of those coefficients in Table 4-3, interpreted as the initial sale retention rate (exponentiation of the intercept coefficients) and annual depreciation rates (exponentiation of the slope coefficients minus one). Models 1 and 2 are identical except for BEVs, which are pooled in Model 1 and separated by Tesla or Non-Tesla in Model 2. Results are consistent with the trends shown in Figure 4-1: the annual depreciation rates (change in retention rate with age) is approximately 9% per year for both CVs and HEVs and steeper for PHEVs and non-Tesla BEVs at approximately 15% per year each. Tesla BEVs depreciate at a slightly lower rate of 12% per year but start at a much higher intercept than any other vehicle type, retaining 98% of their MSRP at one year of age compared to only 70% at one year of age for non-Tesla BEVs. CVs and HEVs both retain approximately 81% of their value at one year old and PHEVs retain slightly more at 87%.

Table 4-3: Estimated First-Year Retention Rates and Annual Depreciation Rates for Different Vehicle Powertrains.

Powertrain	Initial Sale Retention Rate *exp(intercept)*		Annual Depreciation Rate *exp(coefficient) - 1*	
	Model 1	Model 2	Model 1	Model 2
Conventional (CV)	81.08 (0.02)		8.77 (0)	
Hybrid (HEV)	82.14 (0.07)		9.01 (0.02)	
Plug-in Hybrid (PHEV)	86.77 (0.24)		15.34 (0.06)	
Battery Electric (BEV)	72.73 (0.18)	--	13.62 (0.05)	--
Non-Tesla BEV	--	70.27 (0.2)	--	15.4 (0.06)
Tesla BEV	--	98.11 (0.47)	--	12.11 (0.1)
Number of observations:	9,015,324	9,015,324	9,015,324	9,015,324
Adjusted R-Squared:	0.32513	0.331	0.32513	0.331

To understand greater detail in the factors associated with depreciation within each powertrain, we estimate additional models on all cars within each powertrain. Again, we present the transformed coefficients in Table 4-4, with the estimated coefficients presented in Appendix C: Table C-3. These models included interactions effects between the vehicle age and each vehicle model to allow for different depreciation intercepts and rates by model, presented in Appendix C: Table C-4.

Table 4-4: Estimated Effects of Vehicle Model Years and Vehicle Characteristics on Retention Rates, Computed Using Estimated Coefficients in Appendix C: Table C-3. Vehicle models were interacted with age (in years) and are presented in Appendix C:Table C-4.

	Non-Tesla BEV	Tesla BEV	PHEV	HEV	CV
Initial Sale Retention Rate					
exp(intercept)	23.95	102.27	72.68	89.1	104.37
	(0.32)	(2.21)	(1.14)	(0.37)	(0.12)
Difference by Model Year *Reference Level: 2012*					
2013	4.33	10.56	-1.09	-1.88	0.83
	(0.36)	(1.88)	(1.3)	(0.36)	(0.12)
2014	6.33	-11.58	9	-1.57	-1.82
	(0.41)	(1.48)	(1.48)	(0.36)	(0.11)
2015	6.49	-14.64	8.66	-3.59	-5.28
	(0.41)	(1.34)	(1.48)	(0.35)	(0.11)
2016	6.16	-22.66	18.46	0.74	-4.49
	(0.42)	(1.1)	(3.51)	(0.37)	(0.11)
2017	7.73	-26.62	17.78	-0.41	-6.26
	(0.51)	(1.04)	(3.46)	(0.36)	(0.11)
2018	11.23	-47.42	21.42	-1.29	-5.38
	(0.7)	(0.75)	(3.58)	(0.36)	(0.11)
Percent Change in Retention Rate from Addition of...					
...10,000 miles	-5.07	-3.93	-4.49	-4.67	-4.64
	(0.06)	(0.08)	(0.03)	(0.01)	(0)
...10 days on market	-0.37	-0.13	-0.23	-0.28	-0.08
	(0.02)	(0.06)	(0.02)	(0.01)	(0)
...1 cent per mile (operating cost)	0.13	0.38	0.35	1.13	0.08
	(0.06)	(0.13)	(0.06)	(0.02)	(0)
...10 miles driving range	5.58	1.57	-0.8		
	(0.14)	(0.04)	(1.45)		
...$7,500 subsidy in new market	-3.25	-3.41	5.8		
	(0.36)	(0.31)	(0.26)		
Number of Observations:	74,331	22,518	58,915	464,560	8,395,000
Adjusted R-Squared:	0.688	0.604	0.838	0.727	0.587

One of the first notable observations from these models is that while Non-Tesla BEVs and PHEVs have in general depreciated faster than HEVs and CVs historically, this appears to be changing over time. Specifically, we observe significantly higher initial sale retention rates for these powertrains with each new model year, with Non-Tesla BEVs increasing by 11.23% and PHEVs increasing by 21.42% for model year 2018 relative to 2012. The opposite trend has occurred for CVs and HEVs, falling by 5.35% and 1.29%, respectively, for model year 2018 relative to model year 2012. Notably, Tesla

BEV retention rates have fallen steeply over this period, with the 2018 model year initial sale retention rate being 47.42% lower than that of the 2012 model year. This is not surprising considering that Tesla lowered its pricing in the new market multiple times during the period when the vehicles in our sample were listed in the resale market. These results suggest a more competitive market is emerging across alternative powertrain vehicles, with PHEVs and Non-Tesla BEVs performing increasingly better in terms of value retention in the resale market (though not yet quite as good as HEVs and CVs). Figure 4-2 illustrates this trend, comparing the predicted two-year old retention rates of a 2018 versus 2014 model year vehicle for every vehicle model in our analyses. Two-year retention rates were used here due to the available data and model years, and this option allowed us to examine more different model years and observations given data limitations. While not all vehicle models were available in both years, for those that were we observe a steep increase in the predicted retention rates of PHEVs and Non-Tesla BEVs and a steep decline for Tesla BEVs.

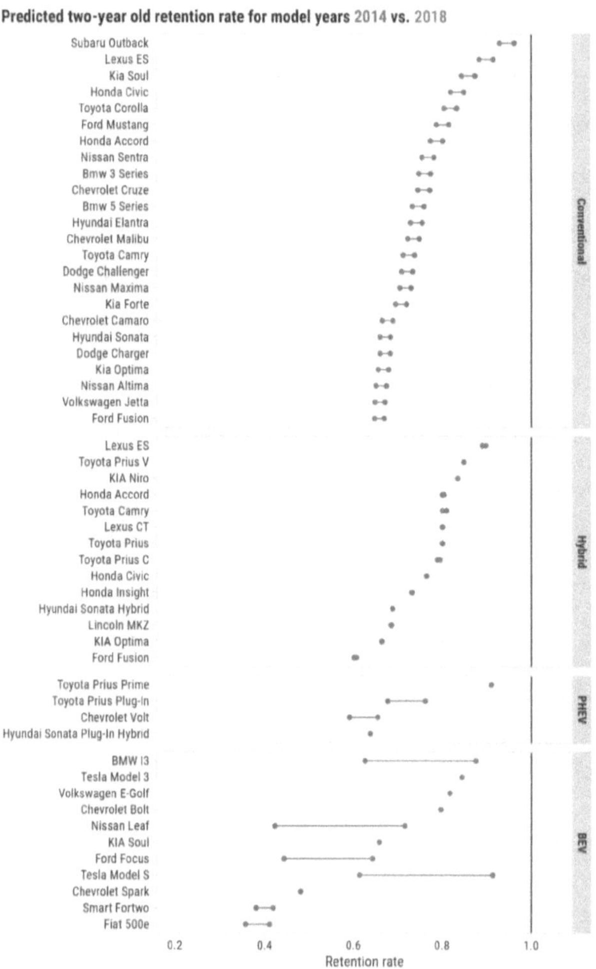

Figure 4-2: Predicted Two-Year-Old Retention Rates for All Vehicle Models, Comparing Model Years 2014 and 2018.
Model year 2018 HEVs, CVs, and Non-Tesla BEVs are experiencing lower retention rates relative to model year 2014 while PHEVs and Non-Tesla BEVs are experience higher retention rates. Predictions are made using the estimated model coefficients in Appendix C: Table C-4

Table 4-4 also shows other important findings regarding how retention rates vary with different vehicle characteristics. We observe that increased mileage has a negative impact on retention rate across all powertrains, with Tesla BEVs having the lowest sensitivity to mileage at 3.9% for every additional 10,000 miles and Non-Tesla BEVs having the highest sensitivity at 5%. We also observe lower retention rates for vehicles that remain on dealership lots longer, suggesting that (as would be expected) dealerships reduce the prices of vehicles that take longer to sell. CV prices fall at lower rates (0.08% for every additional 10 days on the lot) relative to non-CVs, which fall at rates between 0.13% (Tesla BEVs) and 0.37% (Non-Tesla BEVs). We also find that less efficient vehicles (those with higher operating costs) appear to hold their value better than more efficient vehicles (lower operating costs) across all powertrains. While this result may be counterintuitive, it is important to note that more efficient vehicles tend to have higher prices; as a result, higher demand for more affordable (but less efficient) vehicles could lead to less rapid depreciation compared to higher-priced (but more efficient) vehicles.

Table 4-4 shows several important effects specific to PEVs. First, we find that BEVs with higher driving ranges have higher retention rates, though this effect is significantly higher for Non-Tesla BEVs, which tend to have lower ranges in general than Tesla BEVs. For every additional 10 miles of range, Non-Tesla BEVs have 5.58% higher retention rates and Tesla BEVs have 1.57% higher retention rates, all else being equal. The range effect is not statistically significant for PHEVs.

Figure 4-3 shows this relationship, plotting the predicted two-year-old retention rate versus range for multiple model years for select vehicle in our analysis. We only include vehicle models that have at least three available model years between 2014 and

2018 as some models (such as the Tesla Model 3 and Chevrolet Bolt EV) only came out in 2018. Predictions are made using the estimated model coefficients in Appendix C: Table C-4, and the points are the mean two-year-old retention rates across all listings for a given model year-range pair. Here we see the rapid increase in two-year-old retention rates for newer Non-Tesla BEVs, which have larger electric driving ranges. In contrast, the slope of line for the Tesla Model S is negative because the negative model year effect is larger than the positive range effect, which is smaller in magnitude than that of Non-Tesla BEVs. This suggests that there may be a limit to the higher value retention obtained from increased driving ranges.

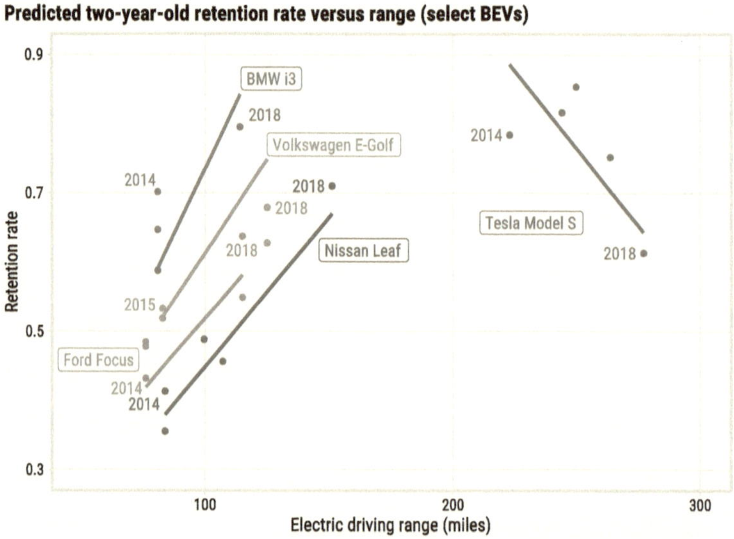

Figure 4-3: Predicted Two-Year-Old Retention Rate Versus Range for Select BEV Models By Model Year (Only Vehicle Models With At Least Three Model Years are Included). Predictions are made using the estimated model coefficients in Appendix C: Table C-4, and points are the mean two-year-old retention rates across all listings for a given model year-range pair.

Finally, we also find an effect from subsidies available in the new vehicle market. A $7,500 subsidy for a new BEV (the maximum federal subsidy for which most BEVs in this time period qualified) had the effect of lowering the retention rates in the resale market of Non-Tesla BEVs by 3.25% and Tesla BEVs by 3.41%. This is expected as lowering the price of a new BEV via a subsidy should also lower prices of earlier versions of the same vehicle model in the resale market. This logic falls from the idea that the upper bound that buyers should be willing to pay for a used vehicle is the MSRP of the new version minus the subsidy value, otherwise they would simply buy the new version and obtain the subsidy (assuming adequate supply). As a result, the retention rates (computed as the listing price divided by the unsubsidized MSRP) should be lower for used vehicles that have an equivalent subsidized model available in the new market compared to those that do not.

Because of this pass-through effect of the new market subsidies into the resale market, we estimate that the total federal subsidies dispensed for new BEVs sold between 2011 and 2019 resulted in $255 million in indirect subsidies to the resale market (see Figure 4-4). While relatively small in magnitude compared to the $8.7 billion in new subsidies, our estimate is still important to note as it is an additional benefit to consumers that comes at no additional cost to the government.

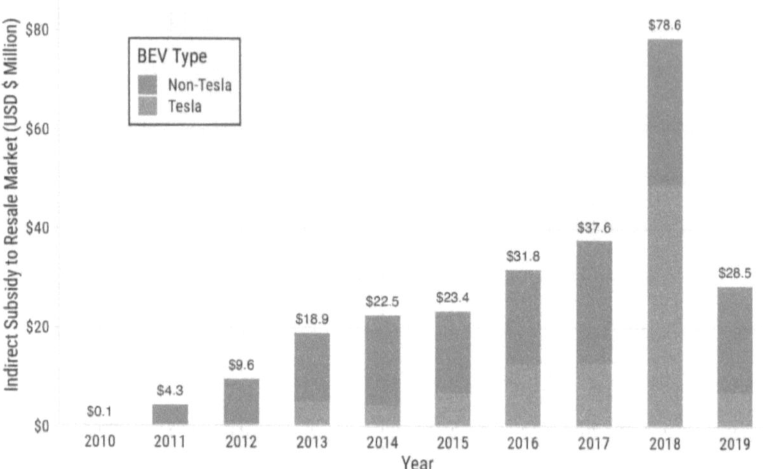

Figure 4-4: Annual Total Indirect Subsidy to the Resale Market from Price Reductions Due to Federal Subsidies for New PEVs.

Although our data extends to vehicles listed up until March of 2022, we chose to censor the data used for all modeling to the end of 2019 due to the supply disruptions in the automotive market that resulted from the outbreak of the COVID19 pandemic in early 2020. These disruptions led to significant increases in vehicle listing prices, with prices for some used vehicles rising higher than the MSRP in the new market since many new vehicles were unavailable for lengthy periods of time. This rise in pricing made models that relied on assumptions of exponential decline unreliable, hence why we omitted the data from our models. Nonetheless, descriptions of the post-pandemic listing prices are still informative. Figure 4-5 shows the mean listing prices for CVs, Tesla BEVs, and Non-Tesla BEVs. As of March 2022, mean inflation-adjusted prices are 37%, 39%, and

3% higher compared to in January 2020 for CVs, Non-Tesla BEVs, and Tesla BEVs, respectively. In combination with the overall trend of increased retention rates for Non-Tesla BEVs, this result suggests that, at least in the short term, the resale market may no longer be a place for buyers to find an affordable BEV, adding an additional barrier to more equitable PEV adoption.

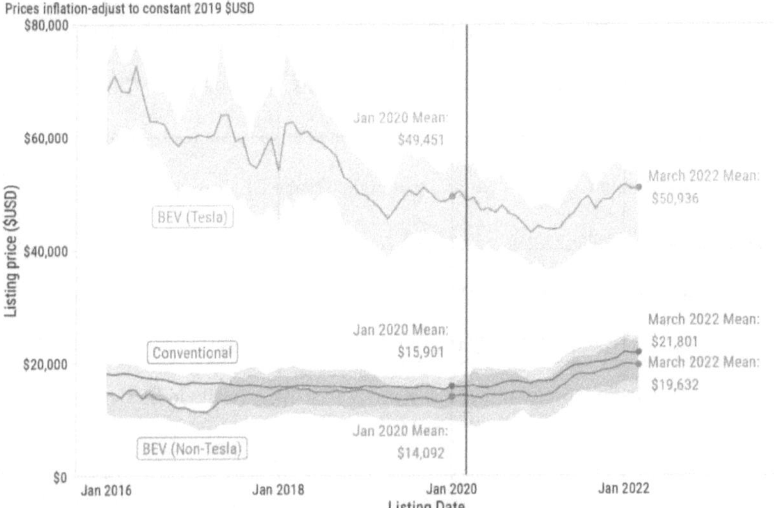

Figure 4-5: Summary of Resale Market Prices for CVs (Gray), Tesla BEVs (Blue), and Non-Tesla BEVs (green).
Solid lines are mean prices and bands show the interquartile range of prices. The vertical red line marks March 2020, the start of the COVID19 pandemic.

4.3 Discussion

PEVs are a critical component in a sustainable transportation future, and the resale market will play a critical role in expanding PEV adoption beyond affluent demographics given its greater affordability [123] and size (more than double in annual sales compared to the new vehicle market [124]). In this study, we analyze the retention rates of different cars listed in the resale market, comparing result across different powertrains. We find in general that PEVs do not hold their value as well as CVs as they age, which is consistent with much of the prior literature on this topic [103], [109]–[111], [114].

However, we also find evidence that this is changing over time, with newer PEV models holding higher retention rates that are approaching those of many CVs. This trend is particularly significant given the rapid advancements in PEV technology, notably in driving range. In our dataset, the mean BEV range grew 76% from 86 to 151 miles from 2012 to 2018. This evolution not only addresses range anxiety but also potentially mitigates resale anxiety [101] as we find that higher-range BEVs have significantly higher retention rates than lower-range BEVs.

Mileage continues to be a critical determinant of resale value across all powertrains, with its impact slightly more pronounced in non-Tesla BEVs. This finding underscores the importance of mileage as a proxy for battery health in BEVs, a key consideration for potential buyers in the used market.

Interestingly, our analysis reveals that subsidies in the new BEV market indirectly influence the resale market, effectively lowering used BEV prices. Our results suggest that a $7,500 subsidy for a new BEV translates to approximately a 3% reduction in retention rates for the same vehicle model in the resale market. While this phenomenon aids in making used BEVs more accessible, it simultaneously could fuel resale anxiety among new BEV purchasers. This dynamic presents a complex scenario for policymakers, especially considering the recent introduction of a used PEV tax credit.

The long-term effects of such incentives on PEV resale values warrant further investigation.

The COVID-19 pandemic has significantly impacted the vehicle market by reducing affordability across all powertrains. Recognizing and adapting to these shifts is imperative in maintaining PEV affordability and, consequently, in promoting broader adoption. The evolving landscape of PEV retention rates, particularly among non-Tesla models, is noteworthy. However, this also introduces a new dynamic where the affordability of subsequent vehicles diminishes due to elevated pricing and retention rates. In the post-pandemic era, the surge in used vehicle prices has somewhat eroded the affordability that once characterized the used PEV market.

Our study is not without limitations. The exclusion of data post-2019 due to the COVID-19 pandemic's impact on the market introduces a temporal boundary to our findings. This exclusion means that our analysis does not account for the potentially lasting effects of the pandemic on consumer behavior and market dynamics, which could have significant implications for the PEV resale market. Also, this significantly limits the BEV models available for analysis, excluding newer BEV SUVs that have been introduced in the past couple years thus leaving out the largest and most popular bodystyle segment in the US from our results. Updating these findings using more recent listings and models would expand the breadth of results here. Additionally, our reliance on listing prices as a proxy for actual transaction prices may not fully capture the nuances of final sale negotiations and discounts. If there are systematic differences in listing and transaction prices by powertrain, then our comparative analysis across powertrains may have slight discrepancies with respect to true differences in retained value.

Nonetheless, the retention rates presented in this study should accurately reflect what consumers would have observed at dealerships prior to making a purchase decision. Our study also focuses predominantly on the US market, which may limit the generalizability of our findings to global markets where different economic conditions, consumer preferences, and policy frameworks prevail. The PEV market dynamics in other regions could offer contrasting insights, especially in areas with different levels of infrastructure development, consumer awareness, overall PEV adoption, and government incentives for PEVs. Our analysis also does not extensively delve into the impact of brand perception and consumer loyalty, particularly concerning Tesla and other emerging PEV manufacturers. The strength and reputation of a brand can significantly influence resale values, and as the market continues to evolve, shifts in consumer perception could markedly alter the landscape of PEV value retention.

Finally, the rapidly evolving nature of PEV technology, particularly advancements in battery efficiency and lifespan, could alter the depreciation patterns in ways not captured in our current dataset. As newer models with improved technology enter the resale market, their retention rates could differ significantly from the trends observed in our study. For these reasons, we caution modelers from extrapolating these trends too far into the future. Continuous monitoring and analysis will be essential to fully understand the long-term trends and implications for PEV adoption and sustainability in the transportation sector.

Acknowledgements

The authors would like to thank marketcheck.com for providing access to the data used in this study. This study was supported by a grant from the Department of Energy.

Chapter 5: Lessons from PEV Research and Adoption of Innovative Technology Theory: What's Next for PEV Adoption in the US?

A more sustainable future is being built today from technological innovations across multiple industries. As the highest greenhouse gas emitter in the US [1], the transportation industry urgently needs these innovations. Plug-in electric vehicles (PEVs), especially for passengers cars, is one of the most salient and feasible solutions for long-term reduction in vehicle greenhouse gases [125]; however, adoption has remained low, especially in the US [126]. To understand how to accelerate PEV adoption, this book explored three different aspects of consumer considerations for purchasing a PEV.

From this research, we have gained valuable insight into how different areas of consumer consideration and adoption of new innovations might impact and deliver further diffusion of this new technology. These studies together connect back to PEV adoption by contributing insights into the various phases along the innovation-diffusion curve. Rogers' Diffusion of Innovation curve, built from the Bass Model, shows that there are many phases in consumer adoption of an innovative technology. These three studies aimed to yield insights along different stages of PEV adoption in order to create a holistic book on the progression of PEVs.

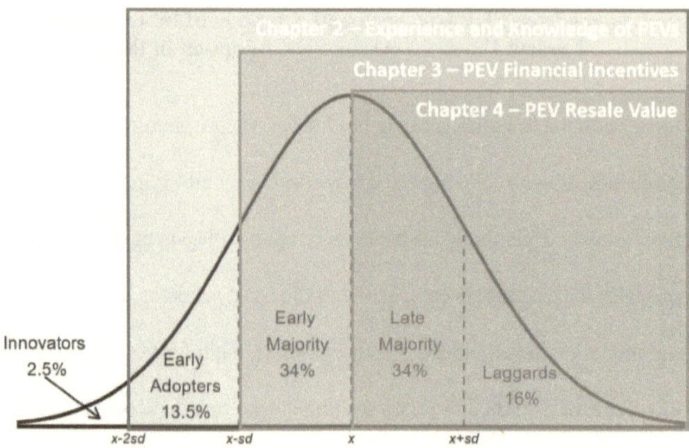

Figure 5-1: Innovation Adoption Curve Overlaid with book Research Chapters and Consumer Impact.
Figure is based on Rogers innovation adoption curve and classifications [9] and recreated by author.

These insights come at a potentially critical point in US PEV adoption as the technology looks to break out past early adoption phase into the mainstream, which potentially means crossing the "chasm" – the transition from "technologists" to "pragmatists" where additional effort is needed to push from early adopters with openness to innovation adoption to the early majority and beyond, which tend to be motivated by more practical measures [127]. Currently, the US PEV market is experiencing some issues looking to bridge this gap and move forward. PEV sales in 2023 slowed compared to growth in 2022, with sales up 46% in 2023 vs. 65% in 2022 [128]. Also, new PEV uncertainties emerged, with one specific example being PEV ownership in the winter and their performance in cold temperatures [129]. And finally PEVs remain a costly endeavor for consumers and, despite price decreases in 2023, the average transaction price of PEVs (roughly \$55k in Jan 2024) remains in line with luxury cars and pickup trucks [130].

Diffusion of an innovation is a highly individual and behavioral phenomenon and can be explored through the Theory of Planned Behavior, which provides some additional insights as it relates to conceptualizing the findings of this book [15]. To create intention to adopt a PEV, three factors contribute to the decision towards intent. First, consumers' attitudes towards the behavior, which we have addressed throughout this book, identify various ways the general perception of PEVs could be improved. For example, as we saw in Chapter 2, allowing more opportunities for the general population to experience PEVs (even for relatively short periods) can create a more positive consideration of PEVs. Next, the subjective norm—or the social standard—for a behavior also contributes to intention. In this book, we saw a few different examples of opportunities to generate a more favorable societal norm when it comes to PEV adoption. Throughout the research, we demonstrated that PEV knowledge was lacking and, with the proper attention, governments at different levels could focus on promotion and education around EVs in order to establish a more societally positive depiction. Finally, perceived behavioral control—or perceived ease of executing on adopting a PEV for our purposes—is the third factor in predicting intention, and this is a key factor addressed throughout this book in the form of equity implications. This research shows there are inequities in PEV consideration and adoption today, including affordability issues, inequitable subsidization, and lack of access.

Each chapter in this book has provided clear contributions to the research of PEV adoption, with analysis into three distinct and significant contributors to technology adoption. In Chapter 2, we showed how short PEV experiences can deliver a positive impact on stated consideration of and recommendation for PEVs. This directly indicates the benefits of PEV experience, (particularly short experiences, which are more realistic and actionable for stakeholders to recreate), on PEV consideration for manufacturers, policy makers, and other PEV stakeholders. There is also concrete evidence that a lack of consumer knowledge around tax credits and refueling of PEVs exists, implying further outreach and education is needed. When considering the theory behind diffusion of technological innovations, this research indicates short experiential opportunities will benefit the early phase of consideration and adoption thus making it a critical component during early adoption and beyond, as well as establishing attitudes, imparting knowledge, and demonstrating the relative advantage, compatibility/complexity, observability and trialability in perceptions of this innovation. Learnings from this study can also be translated to other technologies where an experiential and exposure factor may affect consumers as well as a gap in current knowledge exists – for instance solar panels and seeing a neighbor install a set or a short trial of a geothermal heating system; or a short test drive of emerging hands-free driving systems in a vehicle to gain familiarity with the new technology similar to the research performed in Chapter 2.

Next, in Chapter 3, our results identified issues in current policy structures for PEV financial incentives and how more effective incentive design, based on how consumers value dollars, could be structured. This means offering rebates at the time of sale, which offers additional benefits towards lower income households and would overall improve PEV financial incentive policies effectiveness and equity. Furthermore, this chapter reinforced the lack of knowledge around PEV financial incentive policy, which should encourage further education and outreach for consumers on available PEV incentives. In terms of diffusion theory, this research contributes to the understanding of early majority adopters and beyond as adoption expands to groups with lower social mobility and extends the relative advantage of this innovation. There is also a huge overlap with societal norms and perceived behavior control, as correctly designed subsidies offer more opportunity for access across the entire population and show a larger societal government commitment to this new technology. These conclusions could also assist in incentive design recommendations for adoption of sustainable technologies where cost of adoption is a clear barrier – solar panels and battery storage systems are cost prohibitive for many homeowners so a time of sale rebate would be an important and effective incentive design to consider that could likewise offer an incentive system for furthering adoption.

Finally, in Chapter 4 we provided more robust data analysis on PEV depreciation and residual values in order to deliver a more granular analysis of the PEV resale market. Through retention rate modeling, we demonstrated that while PEVs depreciate faster than their conventional or hybrid counterparts, there has been improvement in recent model years, with increased range providing a crucial improvement over time. Also, our results found new PEV subsidies provide further support to the developing PEV secondary market through downward price pressure. While improvements were being made in PEV retention rates, recent changes to the post-COVID19 automotive market show dramatic price increases even after accounting for inflation and an overall lack of affordability, leaving the secondary market in flux as an affordable option for those considering PEVs.

It is important to note that this analysis focuses on PEV cars and future work should be done to investigate resale value for PEV SUV and truck bodystyles since these are popular vehicles in the US and consumers' interpretation of these PEVs will continue to evolve. This chapter offers insights for innovation diffusion in terms of late majority adopter types, as the secondary PEV market allows those with limited resources and less financial risk to consider PEVs. Also, this assists in the perception of behavioral control, making PEV adoption more attainable via secondary market and an ability to gain relative advantage and observability in perception of PEVs to create a more favorable foundation for PEV adoption. And conclusions here could translate to other higher cost, durable goods as secondary markets will offer ways for certain individuals to engage with these products and technology improvements over time will help improve resale value. And there is a larger secondary market dynamic to consider as we aim more broadly towards a more sustainable future, finding new lives for the goods we consume as a society.

One important thing to note is with any innovative technology, market dynamics are constantly changing and these research studies and findings reflect the time and circumstances in which they were observed. The PEV market continues to evolve, for instance with new PEV models and bodystyles being introduced every year with more technology advancements. Therefore, we should acknowledge that aspects of the research will continue to evolve as well as these updates are brought to consumers. However, throughout the studies here there was consistency in much of the findings across the 3 studies and timeframes, as well as with prior research occurring in earlier years. While this innovative technology will continue to progress, fundamental themes explored here like experience, knowledge, incentives and accessibility remain critical factors in the path to gain consumer acceptance.

Overall, these studies deliver wider contributions to the research community and demonstrate how various levers in innovative technology adoption will continue to push PEVs into the mainstream. The results expand on prior works by investigating novel and up to date insights from unique data sources that provide consumer data supplied to various modeling techniques in order to comprehensively review important consumer relevant factors meant to bridge the gap between current innovations and consumer approval. Throughout these studies, it becomes clear that gaining experience, supported with effective incentives and an affordable used market, supports further PEV adoption, which aligns with innovation technology diffusion theory as well. Also, there are clearly shortcomings in today's PEV market in the US—both technology- and policy-wise—that this book explores more fully. There are some clear recommendations from this research that will help push PEVs forward in the US. Using these three studies holistically, the results comprehensively demonstrate various methods and strategies for more effective and equitable engagement of consumers around PEV adoption.

5.1 Where Do We Go from Here with PEV Adoption in the US?

This book explored the state of PEV adoption and research today and developed further insights into a few specific mechanisms that could help support PEVs. However, there are numerous other topics that cannot be explored in a single book. Furthermore, it is critical to acknowledge the gap between where adoption stands today in the US and the progress needed to continue adoption of this innovative technology in order to achieve a zero-emissions future.

As discussed in earlier sections, prior research has effectively established many barriers, such as range anxiety, high purchase prices, charging infrastructure, lack of basic knowledge, and unwillingness to pay a premium, which spans both PEV technology and policy areas. And while this book has focused on key themes around knowledge, experience, financial incentives, resale value, and equity, spanning both PEV technological performance and policy development, there are clearly many areas to consider to provide a more holistic view into the advancement of PEV adoption in the USA. Therefore, we will take lessons from previous work and this book to discuss the future of PEV adoption in the US, focusing on key insights as they relate to both PEV technology development and PEV policy.

5.2 Technological Advances to PEVs and Future Considerations for Adoption in the US

On one hand, our current understanding of the technological factors affecting PEV adoption are generally well researched and can be thoroughly articulated. At a high level, range anxiety, charging, and high purchase price are consistently cited as key technology barriers of PEV adoption [16], [28]–[34]. While aspects of these factors are being addressed through technological innovations as PEVs continue to evolve and advance, we can observe that some of these efforts will fall short of consumer expectations. For example, there has been an emergence of recent, higher mileage PEVs: Tesla provides some of the longest range PEVs available with models offering about 300+ miles for their long-range options and Lucid Air was recently introduced with over 500 miles of range (albeit with a price tag of over $100,000) [131]. Considering that just ten years ago PEVs were being introduced with around 100 miles of range, the range developments have been impressive. However, even with these significant improvements, the 2021 median range for electric vehicles is 234 miles, while the median range for

gasoline vehicles is 403 miles. [132]

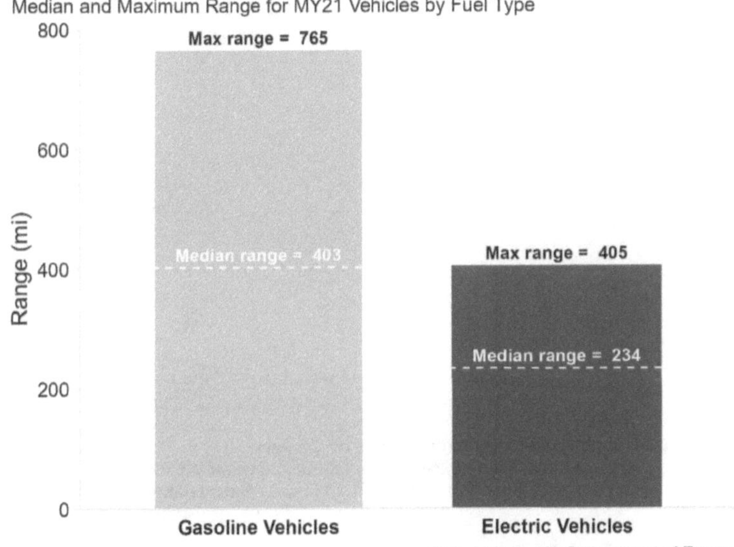

Figure 5-2: Median and Maximum Range for MY21 Vehicles by Fuel Type.
Despite PEV Range and Battery Technology Improvements Over the Years, PEVs, on Average, Still Fall Well Short of ICE Vehicle Range. Figure recreated by author utilizing data from US Department of Energy [132].

Additionally, in the US automotive utilization patterns differ from those in other

countries in that vehicles are typically needed for the majority of day-to-day tasks, and

the distance Americans expect to travel on a regular basis is higher. As seen in the 2022

Global Automotive Consumer Study by Deloitte, US consumers' expectation of all-

electric driving range is 518 miles, 100+ miles higher than the average range expectation

of all countries surveyed [133]. This type of range is out of reach for almost all PEVs

today and achieving these types of ranges will affect battery size and price, leaving this

range at an affordable price currently unfeasible by a wide margin.

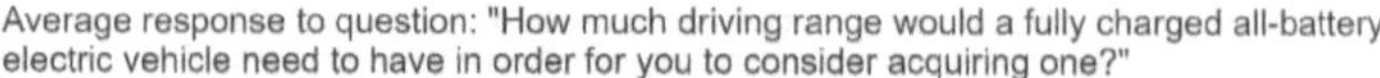

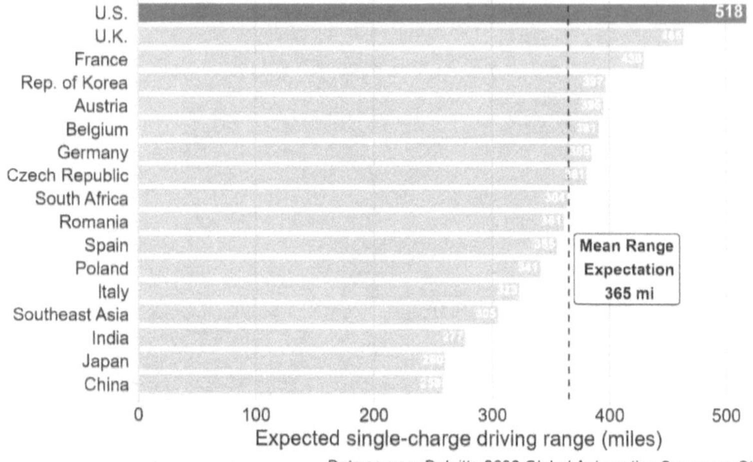

Figure 5-3: Consumer BEV Range Expectations by Country.
Regionality is a Key Input into Range Expectations with US Consumers Expecting 500+ miles of BEV Range. Figure recreated by author with data from Deloitte Global Automotive Consumer Study [133].

Enhancing the energy capacity of PEVs is a key R&D initiative. With explorations into solid-state batteries, different and more efficient mineral compounds (i.e. sodium-ion), and a reimagination of anode and cathode materials, we will likely see longer-distance and more efficient and affordable battery technology in the future [134]. The US has seen an influx of local EV manufacturing and battery production, with US battery manufacturers like AESC, Gotion, LG, Northvolt, our Next Energy, Panasonic, SK Battery America and Redwood materials increasing local R&D and production of EV batteries, materials, cathode and anodes, and recycling for both PEVs and energy systems [135]. However, just to achieve PEV 2030 targets in the US, PEV battery production will require $10-73 billion in total investment [136]. PEV battery improvements are clearly expensive and require long development cycles, which is directly at odds with the policy requirements for increasing PEV sales over the next decades. With this, experience will be a critical element to help give consumers understanding of PEVs; governmental support will also be an essential element to continuing promotion of advancements in PEV driving range technology.

Charging is another significant technological change compared to ICEs and is frequently cited as a barrier to PEV adoption. Compared to an ICE, charging is a much more time-consuming and unfamiliar process [137]. When first launched, PEVs introduced a massive shift in the "refueling" process, including different levels of charging, 50kW fast charging capacity, different plug types and much confusion for consumers [138]. In the past decade, the progression of this technology continues with PEVs available now with advanced 800V charging architecture and 350 kW fast charging capabilities. With these advancements, certain vehicles like the Hyundai IONIQ 5 can fast charge in 18 minutes in ideal conditions [139]. While PEV charging time has progressed to a more reasonable timeframe, overall, it still requires a substantial amount of time as compared to ICEs. Charging time innovation continues—with researchers investigating different charging electrochemistries, electrolytes, solvents and additives, or battery algorithms—however these efforts remain in laboratories currently and there are some critical roadblocks to mainstream utilization, particularly industrialization and production, electrical grid capabilities, and mineral supply chain [140]. Further exploration and research are needed to make more advances in reducing charging time, which will require more time and investment to deliver.

The location, amount, and reliability of Electric Vehicle Supply Equipment (EVSE) for charging is another critical area of future development for the US. Consumers question where charging will occur: home charging is subject to the individual housing circumstances of each household while also potentially requiring further financial investment and fast-charging stations are still further developing. Currently, the US government is taking on an ambitious plan to expand access to chargers by 2030 for $7.5 billion and potentially 500,000 new chargers [141], [142]. Currently, one major complication in EVSE technology is in the US there are different EV charging plug technologies available simultaneously, making universal EVSEs impossible and restricting and confusing customers [143]. Increasing charging locations and total number of EVSE equipment is a focus for the future, however research has found today's public fast charging access is distributed inequitably, making this another important accessibility point of contention for the future [144]. And while PEV charging technology has progressed significantly, there are still many open questions for customers. Applying some of the lessons from this book, addressing lack of experience and knowledge around EV "refueling" and adding a federal ESVE financial subsidy for equipment and installation would help address the technological shortcomings. On the technology side, further innovations must be expected, including a universal plug for PEVs sold in the US, increased access and equitability in EVSE charging locations, expansion of vehicle-to-X (V2X) capabilities, and development of multi-household and workplace charging stations on a mass scale.

The costs associated with PEVs is another oft-cited issue for PEV adoption. High purchase price is one of the aspects in PEV cost analysis research that is specifically cited as a factor constraining PEV adoption [4], [16], [28], [29]. Between federal and state differentiation and variation of incentives, a convoluted system has developed in the US around PEV incentives as a mechanism to increase consumer adoption. The most expensive piece of technology on a PEV is the battery [145]. From their inception, PEV prices have decreased over time due to further industrialization yielding lower-cost batteries and manufacturing. Since PEV development started in 2008, the estimated cost of a PEV battery has dropped 89% to $153/kWh [146].

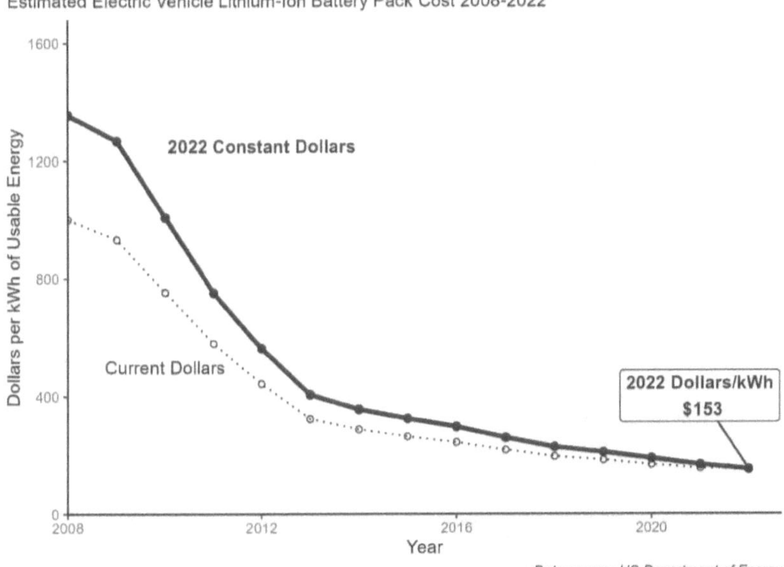

Figure 5-4: Estimated Electric Vehicle Lithium-Ion Battery Pack Costs from 2008-2022. Figure recreated by author with data from US Department of Energy [146].

However, this drastic change stands in contrast to the change in where electric vehicles transact today, despite the battery cost decreases already delivered over the past decade. In December 2022, the average transaction price for of an electric vehicle was $61,448 (before incentives) approximately equivalent to an entry-level luxury vehicle [147]. The Tesla Model 3 set high price expectations with a promised entry price of

$35,000 when it was launched in 2018, however this version was never really sold in high volume, and currently the cheapest Tesla on their website is $40,630 excluding taxes [148]. While recognizing the COVID-related impact to supply chain shortages and price increases across the entire automotive industry recently, PEVs currently transact at a level that makes them unrealistic for most households. Policy could address this by giving automakers and battery suppliers incentives for targeting lower price points for PEVs and adding further consumer incentives to the used PEV market, which this book explored as a more affordable option. In regards to PEV technology costs, battery prices still need to decrease further in order to attract more consumers, which will necessitate much more investment across the entire battery supply chain [149].

Technology advances are an important part of how PEVs continue their path to adoption. The past decade has delivered exponential advances in the range, price and charging delivered to consumers and increased PEV adoption over the past decade to today. If you compare PEVs produced ten years ago to those today the level of innovation is evident, but the rate at which these technological innovations deliver further advances will slow as the technology matures and improvements become more incremental [150]. The next significant phase of technological innovations will require billions of dollars in investments and come with an uncertain or long lead time; solid state batteries, 5-minute PEV charging, and other innovations are still under development, (without any specific timeline for consumer introduction), which means for the impending years it is important to consider what other mechanisms outside the technological advances should be implemented in order to continue driving consumer adoption. This is why applying the lessons from this book to policy mechanisms for PEV adoption in the US will be so critical in the nearer term to deliver more zero-emissions vehicles.

5.3 Policy Developments and Future Mechanisms for Driving PEV Adoption

To further encourage PEV adoption—while technological attributes like range and charging have time to further improve and gain consumer acceptance—there are numerous policy mechanisms to consider. Policy can also be used to address equity issues in innovation diffusion by bridging the known gap between socioeconomic statuses.

In the US currently, the main policy drivers for increasing PEV adoption are prescriptive, through regulatory targets and enforcement aimed at OEM sales percentages. Most notably, the state of California and CARB approved updates to its ZEV mandate via Advanced Clean Cars II to deliver Governor Newson's executive order target of 100% of in-state sales of new passenger cars and trucks to be zero-emission by 2035 [151]. Penalties for OEM non-compliance of current ZEV mandate include significant civil penalties for every vehicle out of compliance [152]. This has also been adopted by 12 other states (and DC) explicitly planning to follow Advanced Clean Cars II and also potentially for other states that follow today's CA ZEV mandate regulations under Section 177 of the Clean Air Act [153]. Also, at the federal level President Biden signed an executive order that sets a target of 50% of all new vehicles sold in the US to be zero-emission vehicles, though there are currently no consequences associated with this executive order target [154]. These actions are helpful in providing regulatory enforcement and establishing timeframes and PEV adoption rates for OEMs to target in order to avoid negative consequences, both legal and PR. However, they don't actually generate increased consumer demand of PEVs and don't address some of the key underlying issues as identified in this book such as lack of experience and knowledge of PEVs, and affordability and equity.

Policies aimed at consumers for amplifying PEV adoption tend to focus on incentives due to their demonstrated effectiveness. In Norway, the majority of new car sales are PEVs due to strong, positive incentives for PEVs, (exempt from 25% VAT tax, no road traffic insurance tax, 50% toll pricing, and additional benefits depending on municipalities), combined with negative consequences for ICEs, (25% VAT tax, 20% carbon tax, smaller weight/NOx taxes, and a car scrapping fee) [32]. Research and real-world examples conclusively demonstrate financial incentives specifically are effective in increasing PEV sales [24], [45], [76], [155]–[160]. Currently in the US we fall short in these policies, especially financial incentives, which are needed in order to support the near-exponential growth projected over the next decades. Nationally, The Energy Policy Act of 2005, the Energy Independence and Security Act of 2007, and the Energy Improvement and Extension Act of 2008 supported the initial creation of the current federal tax credit for PEVs, with a maximum subsidy of $7,500 [161]. Beyond the federal tax credit, some states offer PEV financial incentives (subject to budget availability and other factors) that can take many different forms (frequently tax/excise credits) [122]. So, for federal and many state incentives, consumers still pay the full price for the PEV at the dealer (and higher sales tax etc.) and then must wait until filing taxes to get the credit, subject to individual tax circumstances. In fact, delving further into the impact of individual tax circumstances, households with children and lower incomes are less eligible to receive full PEV tax credit [162]. This has led to consumers devaluing a tax credit incentive as compared to a rebate that could be delivered at time of sale, as noted by this book. It has also created an equity issue where lower-income households 1) are not eligible for the full tax credit amount currently based on today's tax system; 2) more heavily devalue a tax credit vs. time of sale rebate than those with higher incomes; and 3) may not utilize it all, because it does not apply to used PEVs.

Updates to the federal PEV tax credit policy have been introduced in 2023 via the Inflation Reduction Act (IRA), which implemented many changes to sustainable policies; however, the revamped PEV tax credit is subject to new local production, battery and mineral thresholds that have shifted its intent as well as made it confusing to consumers. One key enhancement that has come via the IRA for 2024 and beyond is the ability to apply the federal tax credit at time of sale and while a notable improvement over prior policy design, research from this book found time of sale tax credits remained less valuable to consumers vs. rebates and in addition there are far less vehicles that qualify for this version of the tax credit (20 PEVs, including 14 BEVs as of March 2024 [163]), so it does not substantially tackle the lack of effectiveness and equity issues presented by the previous federal tax credit [44].

PEV rebates have also been discussed, with some states that are focused on ZEV sales implementing time-of-sale rebates, but due to the constraints of state budgets and operations these rebates can exhaust their funding and lapse throughout the year and can require an application/approval process, thus still not truly being delivered at time of sale and also are subject to further additional constraints like maximum vehicle MSRP limits and minimum electric range requirements [122] , [164]. In addition, some states have reduced or repealed prior PEV incentives before adoption had a chance to substantially grow. Figure 5-5 illustrates a case study in Georgia where a $5,000 tax credit was abruptly dismantled, causing an immediate drop off in alternative fuel license plates and clearly demonstrating the importance of financial incentives to PEV adoption [165]–[167]. To really be effective in the policy space for PEV consumer adoption, there needs to be a holistic, consistent national approach to financial incentives that offer PEV time-of-sale rebates with an emphasis on creating an effective and equitable implementation subject to less constraints for simplification to consumers. This could be done with specific secondary policies aimed at lower-income PEV adoption, assistance in receiving loans for vehicle purchases, and government-funded access to charging in multi-home dwellings. Also, as demonstrated throughout this book, there is a lack of knowledge surrounding current federal PEV incentives that needs to be addressed through further consumer outreach, education, and campaigning.

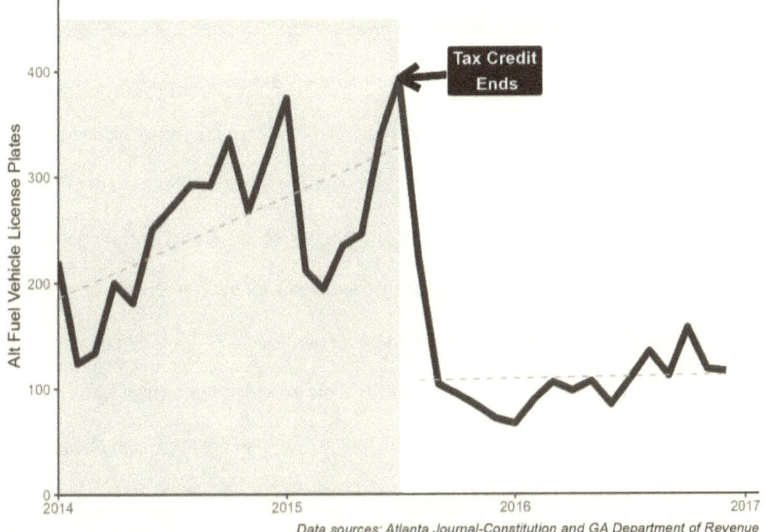

Figure 5-5: Alternative Fuel Vehicle License Plates Issued in Georgia During and After Availability of State Tax Credit.
Figure recreated by author utilizing digitized data from Atlantic Journal Constitution and GA Department of Revenue [165].

The IRA also introduced a vehicle incentive to be given in the used market, which is an advantage as used PEVs also stand to make a big impact equity-wise, with generally lower transaction prices, and environmentally, with older, less-efficient vehicles being replaced. However, there has been very little done to promote or explain this incentive so far and again, this is still implemented as a tax credit with specific stipulations on qualifying vehicles, number of owners, and more [168]. More can be done in this space to truly offer a value PEV to those consumers who need it most and create more equity. Advancing these types of policies along with regulatory targets and other incentives would offer a more substantive path for US PEV adoption.

One critical thing to note is while these important PEV policy developments over the past few years have made an attempt to promote PEV adoption in the US, there is emerging research showing that US PEV consideration and adoption is becoming increasingly politicized. Socio-political attitudes can influence consumer behaviors, and with new environmental innovations begin introduced, political affiliation can affect consumers' intention and adoption of these sustainable technologies [169]. Partisan dynamics have directly impacted PEV consideration and adoption with opinion of PEV car brands (like Tesla, Rivian etc.) and potential to purchase a PEV reflecting a polarized contrast depending on political affiliation [170]. This extends even further with research showing Democrats are significantly more willing to adopt PEVs than Republicans in the US [171]. This presents a pressing obstacle in continuing to grow PEV adoption in the US. This book identifies some key themes that can be extended to address this specific challenge in the coming years. Further education and experience to these communities with messaging focused on PEV benefits, financial motivations to consider PEVs and consistent federal political direction would help demonstrate a societal commitment to PEVs and future sustainable endeavors.

In general, for the US to demonstrate its continued commitment to PEV adoption and push beyond the early adopter and niche technology phase, there needs to be a more holistic, robust, and consistent policy approach. The technology will continue to develop and mature, offering further benefits to consumers, but the near term will require more creative thinking and approaches to policy measures. The federal government recently phased out the only national consumer-facing mechanism for encouraging PEV adoption, a policy developed in 2005-2008, in favor of a complex, multi-step policy that serves to incentivize localized vehicle and battery production over consumer interests; there is still much opportunity in these areas and examples and benchmarks to investigate for increasing PEV adoption. Now, more than ever, we are aware of the negative consequences to the environment and the market has reached a point where high levels of adoption, like those in Norway, are possible. First, rebates as incentives have the best consumer impact and allow a more equitable environment for consumers. There also need to be less administrative details for consumers to consider. Other PEV incentive policy avenues to consider is better incentivization for charging access (both public and private) as well as regionalized non-financial incentives. Depending on region, incentives like HOV access and others can be effective in the right setting [172].

There also needs to be a much higher degree of emphasis on communication, experience, and promotion around PEVs. Right now, creating mandates is not generating the organic consumer demand that is critical to moving towards mass adoption in the US. With any innovative technology there is a possibility of incomplete or discouraged adoption and these next five to ten years will need mass communication and PR campaigns around PEVs illustrating the advantages and benefits that may not be obvious or apparent to adopters. This book has demonstrated a consistent lack of basic knowledge around PEVs; a holistic digital, physical, and media campaign administered at a federal level would kick off a much broader and universal knowledge-sharing imperative for increasing awareness and reducing uncertainty. Mass communication to the public regarding available federal subsidies for new and used vehicles with clear, concise information on their utilization. Short experimental events could be facilitated by different stakeholders, including local governments with PEV fleets and dealers with

PEV products. Promotion of growing charging networks and efforts should be more
heavily publicized and solutions for homeowners, developers, businesses and multifamily
dwellings should be supported, promoted, and mandated with clear strategic and
administrative support. Borrowing from Rogers' attributes of innovations, exploring
more trialability for PEV ownership would help reduce this uncertainty and increase the
rate of adoption. A 3-month trial program or subscription service promoted by local
governments, automakers or dealers would also offer the ability to test a PEV and offer
the flexibility some adopters are looking for in their decision process. Finally,
demonstrating a push for PEVs hand-in-hand with other sustainable technologies (solar
panels, battery packs, vehicle to grid charging technology) would give adopters holistic
exposure to a wider set of sustainable concepts and technologies that are critical to
creating and adopting more sustainable technologies and innovations for our future.

Chapter 6: Conclusion and Broader Theoretical and Policy Implications

This book delivers insights into accelerating PEV adoption in the US through three applied research studies. Each study provides specific contributions, including the benefit of short experiences for increasing PEV consideration, PEV rebates as effective and equitable financial incentives, and improvements to PEV resale value over time. While these studies provide critical observations into consumers' perception of PEVs, there is also more to gain by considering the overall theoretical and policy implications from this book.

There are many common themes between this book and prior research on new innovation adoption in general. When considering the broader theoretical impacts from this book, the context of PEV adoption in the US at the time these studies were conducted is a key consideration. According to Rogers's technology adoption curve, most prior research (including the research from this book) occurred during the "early innovators" / "early adopters" phase of innovation adoption. PEVs have currently not yet crossed the "chasm" into the early majority phase. With the goal of majority adoption of PEVs in mind, this book demonstrates that there is potentially greater importance on the theoretical individual inputs as highlighted by Ajzen and Rogers at this early adoption phase [9], [15]. In other words, to break out past early adoption phases, this research suggests early phase individual inputs like experience, knowledge, incentives, and accessibility are more important in shifting attitudes, perceptions, and building knowledge, as is described in prior theoretical works. There is also a clear reinforcement between the relationship of these individual input traits to the overall consideration or intention to adopt an innovative technology. This again emphasizes the further benefit of considering these inputs with higher significance in trying to bring an innovative technology past early adopter phase into a majority adoption and transition into the dominant technology.

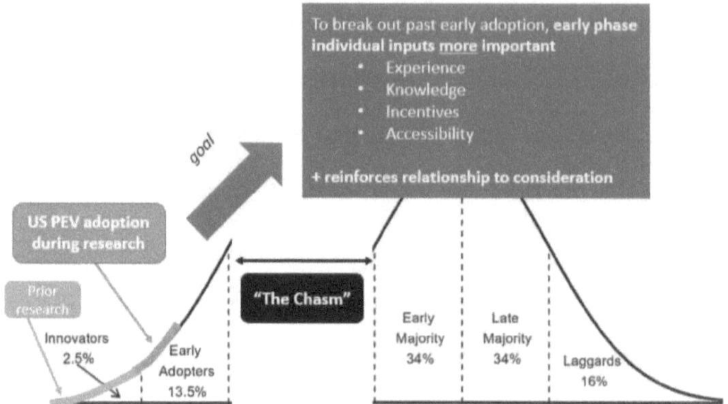

Figure 6-1: book and context of adoption demostrates increased important of early phase individual inputs
Figure created by author utilizing concepts from Ajzen, Rogers and Moore [9], [15], [127]

With this theoretical construct in mind, the results of the three studies in this book can be translated into actionable policy for the many stakeholders focused on increasing PEV adoption beyond the early adopter phase. It is also important to note that these policy implications may not apply in later stages of PEV adoption (a time when different policies may be needed). From the research in this book, I summarize two main policy suggestions: 1) augment current PEV subsidies from tax credits to immediate rebates for both new and the used market, and 2) support both direct experience and education campaigns to increase consumer awareness and knowledge about PEV technology.

To date, the vast majority of government incentives are focused on tax credits for new PEV purchases, with little to no allocation towards experience and / or education campaigns. This book highlights that both experience and education matter, and creating a more comprehensive suite of policies that integrates this could yield more positive adoption results. Such policies could be adopted at the state and federal levels of government.

From a state perspective, California and 12 other states (and DC) are currently planning to follow Advanced Clean Cars II to a target of 100% ZEV new car sales by 2035. These states in particular should consider revised policies to deliver on the ZEV sales targets in a short amount of time, and doing so may not be prohibitively expensive. Virginia makes for an interesting case study. While the state has committed to Advanced Clean Cars II targets, it historically has not implemented significant policy towards increasing PEV adoption. With a state budget forecasted to increase in general fund revenue of $5B [173], [174] over the next couple of years, there is opportunity to earmark a portion of these funds to support PEV adoption. Namely, allocating funding for both a new and used vehicle state rebate, a ride-and-drive experience campaign, and a PEV education campaign would harness the learnings from this book into a comprehensive set of policies aimed at delivering the target growth in PEV adoption. It is important to note that these policies would only consume a small percentage of the $5B increased revenue projections. For example, 6% of the budget ($262M) could be allocated as follows: a $2,500 new PEV rebate for 35% (2026 Clean Air II target ZEV % sales) of roughly 300,000 new vehicles in VA [175]; $3M for ride-and-drive experiences at $10,000 (benchmarked Smart Columbus estimates [176]) per event for biannual events at 133 counties in VA; $2M for various public relations and educational executions (e.g., social media campaigns, mailers, bulletins, community advertising, PR campaigns etc.), and the remaining $33M for used PEV rebates (sufficient for 33,000 used vehicle sales). With just 1% or 2% more, additional policies could be funded, like programs for charging equipment installation and additional PEV rebates for lower-income households. And further activities could occur at the city and county level using existing resources, such as holding ride and drive events with PEVs the city fleet already owns, creating a more holistic outlook for PEVs utilizing the learnings from this book. This book highlights the lack of experience and knowledge that consumers have with respect to PEVs and also the positive impact that improving experience and knowledge can have on PEV adoption considerations. If states like VA are serious about achieving the ZEV adoption targets they have planned, then strategically allocating funding to a suite of well-designed policies that align with consumers' preferences is necessary.

While state policies have an important role to play in PEV adoption in the US, the federal government could have an even larger impact and create more consistent and significant policies to support PEV adoption. The federal government has access to resources much greater than those of individual states and capabilities across the whole US. The most significant change from current federal policies is to transform the PEV tax credit to a rebate at the time of sale. The IRA bill came close to achieving that change as the tax credit can now be applied at the time of sale, but only if dealers are properly registered with the IRS to participate in the program, which means an immediate tax credit is not guaranteed to be available to all customers. Furthermore, very few vehicles currently qualify for the full tax credit due to supply chain sourcing requirements for key

PEV components, such as the battery. In addition, this book found that tax credits at time of sale are still valued less than rebates by roughly $1,100 dollars, so while the timing of the credit is an improvement, it could still be further improved by converting it into a rebate.

For example, if every PEV sold in 2023 received a $7,500 tax credit, that results in almost $10B in tax credits to be refunded to individuals through April 2024. An alternative is to collect that sum in taxes and instead of issue refunds, allocate it towards a PEV rebate fund that is used to pay OEMs or financial institutions based on their PEV sales documentation. These entities already have infrastructure in place to accommodate incentive payments made to dealerships for other sales and discount programs, so this should be technically feasible. Dealers in turn can market the discount as an immediate rebate to consumers, taken off at the point of sale.

Of course, while reallocating funds from a tax credit program to a direct rebate program would benefit the consumer, it might not be politically feasible, so at a minimum a significant rebrand should go on to reposition the tax credit in consumers' minds as a time-of-sale "rebate" rather than as a tax credit. Since there is an ability to use the incentive at time of sale, the mechanics behind the scenes are less relevant and advertising a rebate verbiage has been demonstrated to be more effective in consumers' valuation. In addition, many of the various restrictions on today's PEV tax credit that do not address equity concerns should be removed to lower the barrier of entry for consumers, including battery component/mineral restrictions, location of final assembly and model year and sales price restrictions for used credits. Finally, there needs to be consistency in terms of incentives available to consumers from the federal government for more effective awareness and utilization of this important program.

Beyond financial incentives, there is a need for policies aimed at introducing more consumers to PEVs and facilitating more direct experience with the technology throughout the country. A policy creating PEV ride-and-drive grants for states and a fund for the government to partner with certain organizations to execute their own events would allow more citizens to gain exposure to this new technology. Allocating $4M for 50 states and $50M for federal events translates to $250M to deliver a nation-wide fund to experience a PEV. If that amount of funding were instead applied as an immediate rebate, it would translate to only 33,333 PEVs receiving a subsidy—a small quantity considering the millions of PEVs needed to be sold in the coming decade. It is plausible that instead allocating that funding to an experience campaign may result in even greater overall PEV adoption as the campaign could reach millions of potential PEV buyers.

Likewise, consumer PEV knowledge and awareness is quite low, suggesting a large-scale, federal public education campaign is warranted. Similar efforts have been made in other contexts on similar scales, such as the $250M FDA efforts to prevent youth smoking and a $265M campaign from DHHS encouraging adoption of COVID-19 health measures[177], [178]. A similar widespread campaign aimed at educating consumers on PEVs' benefits and especially available financial incentives is necessary to grow familiarity and consideration of PEVs. Combined, an experience and education campaign could cost around $0.5B—just 0.05% of the $917B non-defense discretionary budget [179]. There could be opportunities to find bipartisan support for such a small amount of funding for these programs, especially if they generated greater economic activity in across all regions of the country. There is also potential for further revenue generating activities to fund these policies, like increasing gasoline fuel taxes or vehicle miles traveled (VMT) taxes, or further payroll/corporate taxes. And none of these proposals consider the costs of future climate change down the road that are difficult to quantify – extreme weather impacts to people and property, pollution health impacts, and more [180]. These simply reallocate or utilize revenue generating activities that already exists to policy makers in order to deliver further PEV adoption in the US.

In conclusion, PEV adoption in the US has made progress, but there are risks to our future planning and achieving environmental goals. By applying the results of this book, these can become opportunities to further promote and educate consumers on PEVs and their benefits, financial and otherwise. Particular attention must be focused on ensuring equitable access to PEVs as historically this has be overlooked and current policies do not holistically address. And fundamentally this research has demonstrated a need to address a general lack of experience and knowledge throughout the US. With more consumer-centric policies and attention to innovative technology adoption theory, there is potential for further PEV growth in the US.